U0282290

[英国]达雷尔·因斯 著 郭铭 译

计算机

牛津通识读本·

The Computer

A Very Short Introduction

译林出版社

图书在版编目（CIP）数据

计算机 /（英）达雷尔・因斯（Darrel Ince）著；郭铭译. -- 南京：译林出版社，2024. 10. --（牛津通识读本）. -- ISBN 978-7-5753-0298-2

Ⅰ. TP3

中国国家版本馆CIP数据核字第2024N66W24号

著作权合同登记号　图字：10-2020-535 号

计算机　[英国] 达雷尔・因斯 / 著　郭　铭 / 译

责任编辑　陈　锐
装帧设计　景秋萍
校　　对　梅　娟
责任印制　董　虎

原文出版　Oxford University Press, 2011
出版发行　译林出版社
地　　址　南京市湖南路 1 号 A 楼
邮　　箱　yilin@yilin.com
网　　址　www.yilin.com
市场热线　025-86633278
排　　版　南京展望文化发展有限公司
印　　刷　江苏凤凰通达印刷有限公司
开　　本　890 毫米 ×1260 毫米　1/32
印　　张　8.5
插　　页　4
版　　次　2024 年 10 月第 1 版
印　　次　2024 年 10 月第 1 次印刷
书　　号　ISBN 978-7-5753-0298-2
定　　价　39.00 元

序 言

王崇骏

算术是关于“数（shù）和数数（shǔ shù）”的技术，前者关乎数的概念与表示，后者关乎数的基本运算与操作。算术既是所有学科的基础，也是所有人都应该掌握的重要技能。从算术诞生之日起，人类就在创造和改良各种计算工具以方便开展各种运算和操作。

两千六百年前中国人发明的算盘，是人类历史上最早的“计算机”之一，算盘通过珠子在竹简或木板上移动的方式来进行加减乘除等运算；到了17世纪，机械装置的计算方式取得了突破性进展，出现了包括帕斯卡计算机在内的各式机械计算机；19世纪初英国数学家查尔斯·巴贝奇发明的差分机，利用齿轮的咬合工艺实现了数学上的差分原理进而实现四则运算，称得上是世界上第一台“通用计算机”的雏形。

1936年，艾伦·图灵发明的图灵机给出了理论上的通用计算机模型，为日后电子计算机的发展提供了坚实的理论基础；1942年，阿塔纳索夫-贝瑞计算机的发明开启了电子计算时代的探索；1946年电子数字积分计算机的诞生则标志着全电子计算机时代的到来，也让冯·诺伊曼体系结构成为现代计算机体系结构的基础。尽管经过七十多年的发展，计算机硬件和软件都发生了巨大变革，但冯·诺伊曼体系结构的核心思想仍然广泛应用于当今主流的计算机系统中。

一般而言，可以从如下几个方面认识和理解计算机。

1. 硅基

冯·诺伊曼结构是一种将程序指令存储器和数据存储器合并在一起的存储器结构。它基于三大基本原则：采用二进制逻辑、程序存储执行，以及计算机由五个部件组成（运算器、控制器、存储器、输入设备、输出设备）。在当代，所有硬件都是基于硅基半导体技术实现的。硅基半导体技术的持续进步，使得这些核心部件不断缩小、集成度不断提高，进而推动计算机性能的不断提升。以上内容，详见本书的第一章和第二章。

2. 微型化

摩尔定律指出，集成电路上可容纳的晶体管数量，每隔大约十八至二十四个月就会翻一番，同时成本也会下降，这一定律在过去几十年中一直得到验证。随着集成电路的微型化和制造工艺的不断进步，单个芯片上的晶体管数量呈现爆炸式增长。微型化趋势使得计算机可以嵌入各种电子和机械设备中，从而更加便携、节能和实用。更多相关内容，请参见本书的第三章。

3. 网络化

1969年，罗伯特·凯恩提出的“网之网”概念直接推动了互联网时代的到来。互联网由众多相互连接的计算机网络组成，在计算机发展史上，这是一个里程碑式的创新。互联网可以被视为一台拥有几乎无限计算能力和数据存储设施的巨大计算机，它彻底改变了人类获取信息和交流的方式，极大地促进了人类社会的信息化进程。更多相关内容，请参见本书的第四章和第七章。

4. 影响

随着计算机技术在各领域的广泛应用和持续赋能，各种重大的安全问题随之而来，其相应的安全防御技术因此引起社会各方的关注。另一方面，计算机对业务的赋能可能会引发业务重整，导致某些行业出现颠覆性创新，甚至使某些技能失去价值，进而引发各种社会变迁。更多相关内容，请参见本书的第五章和第六章。

5. 硅基之后

计算机技术正在持续迭代和进步以突破硅基的局限性，各种新型计算范式，如量子计算、生物计算等，正受到广泛关注和重视。相关的理论研究和实验已经表明，基于这些新型计算范式的下一代计算机正在逐步成为现实。更多关于这些内容的详细介绍，请参见本书的第八章。

总体而言，计算机发展的历程是一部极其曲折而又充满智慧的历史，从最初的简单机械装置，到如今的智能化设备，计算机一步步推动着人类文明的进步。我由衷地希望，也相信，通过阅读这本书，读者能够清晰勾勒出这一发展过程的缩影，并能引发对这项伟大发明的思考与探索，共同见证人类文明的崛起和进步。

目 录

第一章

裸　机

引　言

计算机的主要特点之一是其存储数据的能力。它通过使用0和1的组合来表示字符或数字，以实现数据的存储。每个“字节”是由八个0和1组成的集合；每一个单独的0或1被称为“位”（二进制位）。计算机科学家使用各种术语来描述计算机中的存储。最常见的是千字节、兆字节和千兆字节。千字节是10^3个字节，兆字节是10^6个字节，千兆字节是10^9个字节。

我使用的第一台计算机是埃利奥特803型。1969年，我在大学里修了一门计算机编程课，这门课使用的就是这台计算机。它坐落在一个大约40英尺乘40英尺的房间里，计算机的硬件装在许多金属橱柜里，每个橱柜都几乎可以填满我家里的主卧浴室。你需要把写在特殊纸张上的程序提交给两个穿孔纸带操作员，然后他们准备一个纸带版本的程序。纸带的每一行都包含一系列穿孔点，这些点代表程序的各个字符。

然后，程序被带到计算机室，纸带被一种特殊用途的硬件读取，结果显示在一种叫作邮局电传打字机的设备上：这实际上是一台可以被计算机控制的打字机，它将结果打印在纸上，这些纸张的质量比卫生纸好不了多少。

计算机的存储容量是以字节为单位计算的；埃利奥特803型计算机有128千字节的存储空间，其存储器要占用两个机柜，数据保存在小金属环上。用纸带将数据输入计算机，结果通过纸或产生纸带的打孔机获得。它需要一个操作员来看管，并配有一个操作员可以调节音量的扬声器，以便检查计算机是否正常工作。它没有与外部世界的连接（互联网还没有发明），也没有用于大规模存储的硬盘。第一批埃利奥特803型的最初定价是2.9万英镑，相当于今天的10多万英镑。

当写这一章的时候，我正用一种叫作MP3的便携式音乐设备听莫扎特的音乐。它花了我大约180英镑。它很适合放在我的衬衫口袋里，并且有16 千兆字节的存储空间——比我以前大学里唯一一台计算机的容量有显著的增加。

我在一台被称为上网本的电脑上撰写这本书。上网本是笔记本电脑的精简版，可进行文字处理、电子表格制作、幻灯片演示文稿创建和互联网浏览等。它大约有10英寸长、6英寸宽。它有16千兆字节基于文件的存储空间，可以用于存储文字处理文档，连接到互联网，几乎可以瞬间下载网页，还有1千兆字节的内存用于存储临时数据。

很明显，埃利奥特803型和我现在使用的电脑之间存在着

巨大的差别：临时存储空间、文件存储空间、处理速度、物理尺寸、通信设施以及价格。这种进步证明了硬件工程师的技术和

创造力，他们开发的硅基电路每年都在变得更小、更强大。

现代计算机能力的增长体现在一条被称为“摩尔定律”的定律中。这是硬件公司英特尔的创始人戈登·摩尔在1965年阐述的。它指出，用于实现计算机硬件的硅基电路密度（由此计算机的能力）将每十八个月翻一番。直到我写这篇文章的时候，这条“定律”一直存在。

计算机已经从20世纪50年代和60年代的庞然大物，演变成可以放在你夹克口袋里的技术实体；它已经从最初设想只有大公司才会用于进行工资表和存货管理的电子装置，成长为一种消费电子产品，以及至关重要的商业和工业计算技术工具。平均每栋房子里有30台电脑，它们不仅可以制作文字处理文件和电子表格，还可以操作烤箱、控制电视等媒体设备、调节房间温度。

即使在七十年后，计算机仍然让我们感到惊讶。这会让20世纪50年代的许多计算机科学家感到惊讶，因为他们当时预测世界只需要少量的计算机。计算机也让我吃惊不已：大约二十年前，我认为计算机只是便于阅读研究文献和发送电子邮件；而现在，它已与互联网相结合，创造了一个全球社区，使用视频技术进行通信，分享照片，分享视频，评论新闻，评论书籍和电影。

计算机硬件

本书的一个目的，是描述计算机如何影响我们生活的世界。为了做到这一点，我将描述所涉及的技术和过去十年中出现的各种应用——主要集中于应用程序。

首先介绍计算机的基本体系结构；我将在第二章更详细地描述这个架构。如图1所示。图中的原理图既描述了最早的计算机，也描述了最新的计算机：计算机的基本体系结构在六十年里没有发生任何变化。

每台计算机的核心是一个或多个被称为处理器的硬件单元。处理器控制着计算机的工作。例如，它会处理你在电脑键盘上输入的内容，在屏幕上显示结果，从互联网上获取网页，并进行诸如将两个数字相加之类的计算。它通过"执行"一个计算机程序来实现这一点，该程序详细说明了计算机应该做什么，例如读取一个经过文字处理的文档，更改一些文本，并将其存储到一个文件中。

图1中还显示了存储。数据和程序存储在两个存储区。第一个被称为主存储器，它的特性是，无论存储在那里的是什么，都可以非常迅速地检索出来。主存储器用于存储瞬态数据——例如，一个计算的结果，它是一个更大计算的中间结果——也用于存储正在执行的计算机程序。主存储器中的数据是暂时的——当计算机关闭时，它就会消失。

硬盘存储器，也被称为文件存储或备份存储，包含了一段时间内所需要的数据。存储在这一存储器中的典型实体包括数字数据文件、字处理文档和电子表格。计算机程序在不执行时也存储在这里。

4

主存储器和硬盘存储器之间有许多不同之处。首先是检索时间。有了主存储器，处理器可以在几分之一微秒内检索到一项数据。在文件存储器中，检索时间要长得多；大约是毫秒量级。这样做的原因是，主存储器是硅基的，读取数据所需的一切都是通过电子电路发送的。正如你稍后将看到的，硬盘存储器通常是机械的，并且存储在磁盘的金属表面，其使用机械臂检索

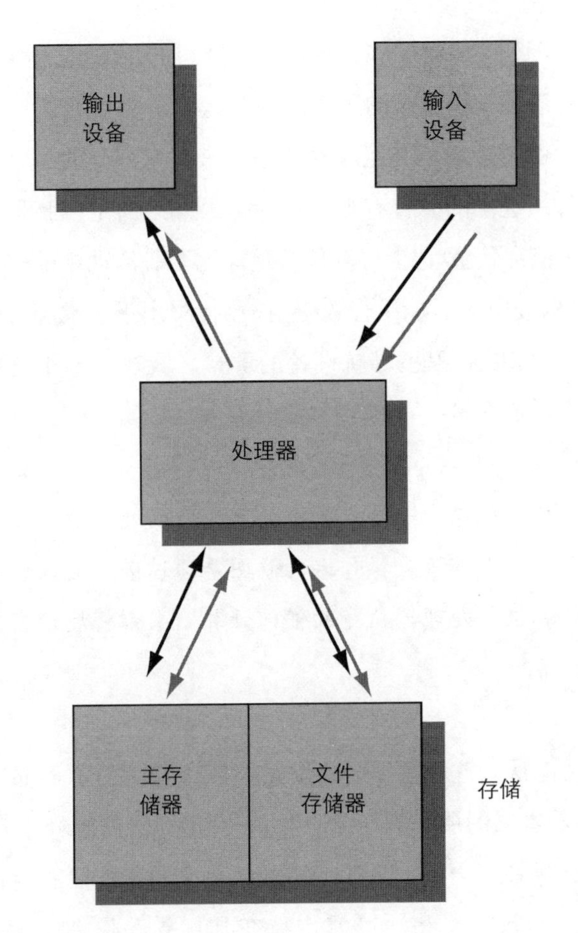

图1　计算机的体系结构

数据。

这两种存储器的另一个区别是，主存储器比文件存储器更昂贵；因此，电脑的主存储器通常比文件存储器少得多（我的笔记本电脑主存储器为3千兆字节，文件存储器为500千兆字节）。计算机的另一组组件是输入设备。它们将用户对计算机执行程序的要求传递给计算机。你最常遇到的两个设备是键盘和鼠标。然而，还有其他一些设备：苹果公司的iPod上的触摸屏与卫星导航系统和核电站的压力监测器是另外两个例子。

计算机的最后一个组成部分是用来显示结果的一个或多个硬件设备。这样的装置有很多种。你最熟悉的是电脑显示器和激光打印机；不过，它也可以包括足球比赛等活动中的广告显示器、飞机驾驶舱中显示飞行数据的控制台、用于生成超市收据的迷你打印机，以及卫星导航设备的屏幕。我在本书中使用的计算机的工作定义是：

> 计算机包含一个或多个处理数据的处理器。处理器连接到数据存储器。操作者的意图是通过若干输入设备传达给计算机。处理器执行的任何计算结果都将展示在若干显示设备上。

你可能认为这种说法既迂腐又不证自明；然而，我希望你能看到，随着本书的展开，对于计算机有许多激进的解释。

在离开本节之前，我们有必要看一看计算机性能增长的另一个指标。奥哈拉和沙德博尔特在其优秀著作《咖啡机里的间谍》中，描述了基于计算机的国际象棋所取得的进步。要想下好国际象棋，你需要预先考虑好一系列的走法，并评估你的对手对这些走法的每一步会如何应对，然后再决定你该如何走出每一步，依此类推。优秀的棋手在头脑中储存大量数据，能够进行快速评估。正因为如此，计算机一直被视为潜在的好棋手。

已经编写的国际象棋程序有效地存储了大量的走法和对

弈，并能很快地对它们进行评估。奥哈拉和沙德博尔特描述道：在1951年，一台计算机只能提前考虑两步棋；在1956年，一台计算机可以在一个更小的棋盘上玩一盘非常有限的国际象棋，但走一步棋要花十二分钟以上。然而，在1997年，一台计算机打败了世界冠军加里·卡斯帕罗夫。这种进步，部分是由于游戏软件技术的改进；而主要原因是计算机的运行速度越来越快。

互联网

计算机不是孤立地运行，大多数都与计算机网络相连。对大多数计算机来说，这将是计算机和通信设备的巨大集合，被称为互联网；然而，它可能是一个控制或监控某些进程的网络，举例来说，一个计算机网络可以保持一架飞机在飞行，或用来监控 7 进出城市的交通流量。

互联网已经对目前计算机的使用方式产生了重大影响；因此，有必要简要地看看它是如何与一台典型的计算机交互的——比如你在家里使用的个人电脑。

互联网是一个计算机网络——严格地说，它是一个连接许多网络的网络。它执行一些功能。首先，它将数据从一台计算机传输到另一台计算机；为了做到这一点，它决定了数据的传输路径。有一个误区是，当你使用互联网进行一些活动时，例如下载一个网页，持有网页的计算机与你的计算机之间的连接是直接的。实际发生的是，互联网通过许多中间计算机算出数据经过的路径，然后再通过它们路由数据。所以，当你看到一个网页在你的电脑上显示时，这个网页可能已经被分成了数据块，每个数据块都经过了一些大陆和一些属于公司、大学、慈善组织和政府组织的中间计算机。

互联网的第二个功能是提高可靠性。也就是说，确保当错误发生时，会有某种形式的恢复过程发生。例如，如果一台中间计算机出现故障，那么互联网的软件将发现这一点，并通过其他计算机重新发送任何故障数据。

互联网的一个主要组成部分是万维网；事实上，“互联网”这个词经常被当作“万维网”的同义词来使用。从现在起我将称它为网络，网络以一种特定的方式使用互联网的数据传输设施：存储和分发网页。网络由许多服务器以及更多的客户机（你的家用个人电脑就是一个客户机）组成。网络服务器通常是指那些比家庭或办公室用的个人电脑功能更强大的计算机。它们将由一些企业维护，并将包含与该企业相关的单个网页；例如，像亚马逊这样的在线书店将为其出售的每一件商品维护网页。

允许用户访问网络的程序被称为浏览器。当你双击桌面上的浏览器图标时，它会向网页发送一条信息，询问你的主页：这是你将看到的第一个页面。互联网的一部分被称为域名系统（通常称为DNS），它会找出页面所在的位置，并将请求路由到保存该页面的网络服务器。然后，网络服务器将页面发送回你的浏览器，显示在你的计算机上。

每当需要另一个页面时，你通常会单击该页面上显示的链接，然后重复这个过程。从概念上讲，发生的事情很简单。然而，它隐藏了大量的细节，包括发现页面存储位置、页面定位、页面发送的网络，读取页面并解释它们应该如何显示的浏览器，以及最终显示页面的浏览器。

我在描述中隐藏了一些细节。例如，我还没有描述视频剪辑和声音文件等其他网络资源是如何处理的。在后面的一章中，我将提供更多的细节。值得一提的是，在这一点上，这些资源在网络上的转移方式与网页的转移方式并没有太大的不同。

互联网是计算机从数据处理机器转变为通用机器的主要原因之一，例如，它可以编辑音乐文件、预测天气、监测病人的生命体征、展示令人惊叹的艺术作品。然而，如果没有一种特定的硬件进步，互联网将成为它自己的影子：这就是宽带。这项技术提供了十五年前我们无法想象的通信速度。大多数互联网用户必须依赖于所谓的拨号设备，它以每秒约56千位的速度传输数据。当你考虑到网页的平均大小大约是400千位时，这意味着需要大约七秒的网页显示在你的浏览器。在20世纪90年代，公司使用专用通信硬件来克服速度的不足。

9

不幸的是，在宽带普及之前，普通用户无法做到这一点。

典型的宽带速度从每秒1兆位到每秒24兆位不等，较低的速度大约是拨号速度的20倍。正如你将在本书后面看到的，这已经改变了家用计算机的角色。

软件和程序

将图1中所示的所有硬件元素结合在一起的黏合剂是计算机程序。例如，当你使用文字处理器时，你正在执行一个计算机程序，它会感应到你输入的内容，显示在某个屏幕上，并在你退出文字处理器时将其存储在文件存储器中。那么，什么是计算机程序呢？

计算机程序很像食谱。如果你看一本食谱，你会看到配料表和一系列的说明，要求你添加一种配料，混合一组配料，并把一组配料放进烤箱。计算机程序非常类似于此：它指示计算机移动数据，执行算术操作，如将一组数字相加，并将数据从一台计算机传输到另一台计算机（通常使用互联网）。然而，食谱和计算机程序之间有两个非常重要的区别。

第一个区别是大小。一个典型的食谱可能包含大约20行文本，而计算机程序包含数百行、数千行甚至数百万行指令。另一个区别是，即使是一个小错误也可能导致程序的灾难性失败。在食谱中，添加四个鸡蛋而不是三个，可能会导致一顿饭的味道或质地有点奇怪；然而，在一个百万级的程序中，输入数字1而不是2很可能会导致重大错误——甚至会阻止程序运行。

我们有各种各样的编程语言。它们被分为高级语言和低级语言。一种高级语言，如Java或C#，具有被翻译为成百上千条计算机指令的指令。低级语言通常与基本的计算机指令有一一对应的关系，通常用于实现需要对诸如化学反应器温度达到临界等事件做出响应的高效程序。

每隔几天，媒体就会报道一个软件项目超出预算或计算机系统严重失败的故事，却很少会将这些故障归因于硬件故障。失败的发生会有两个原因。现有计算机系统出现故障的第一个原因是技术错误，例如测试中没有检测到的编程错误。第二个原因要归于管理上的失败：超出或显著超过预算的项目往往是由于人为因素而发生的，例如，对即将产生的项目资源估计不合理，或者客户改变了他们对系统应该实现的功能的想法。

我个人的观点是，考虑到现代计算机系统的复杂性，项目延迟和开发人员犯错误是不足为奇的。

本书主旨

本书的第一个主题是，硬件的进步如何使计算机能够运用在十年前闻所未闻的领域。放置计算机处理器的电路片可以很容易地用一只手握住，而不是放在一个大的金属柜里。一个容纳16千兆字节数据的记忆棒可以很容易地挂在一个钥匙环上。摩尔定律表明，计算机处理器的计算能力每两年翻一番。你现在可以花不到60英镑买到500千兆字节的硬盘存储空间。这件事情产生了很多影响。首先，在过去的十年里，计算机已经能够做20世纪90年代很少有人梦想的事情，例如，英国电信的“愿景”项目将电视带到了互联网上。第二，计算机硬件尺寸的减小使它们能够实际应用在几年前根本不可能的环境中。

第二个主题是软件开发人员如何利用硬件的进步来生产新的应用。这方面的一个例子就是像苹果公司iPod这样的MP3播放器。iPod和其他设备，如索尼随身听，显然依赖于硬件的进步。然而，他们也依赖于一种基于软件的技术，当其应用于一个声音文件时，将文件压缩到仅占原始大小的10%，而声音质量没有明显下降。

第三个主题是互联网如何将计算机连接在一起，使它们的表现就像一台大型计算机一样。这体现在一个被称为“云计算”的理念中，即数据并非存储在一个本地数据库，而是存储在连接到互联网的许多计算机中，可以由具有相对低级编程技能的互联网用户开发的程序访问。

12

与这一想法相结合的是互联网作为计算机用户可以利用的巨大数据资源。这包括由英国政府和美国政府的数据官网项目发布的数据，也包括由互联网用户直接或间接提供的数据。例如，有些网站可以让你回到你的城镇或乡村，并检查你的邻居正在体验的宽带速度，这些网站包含的数据是由该网站的用户提供的。

第四个主题是互联网如何提供了只有专业人士才能使用的创意设施。例如，计算机硬件的进步、软件的进步，以及用于制作摄像机的技术的进步，意味着任何人都可以成为电影导演，并在互联网上展示他们的成果。计算机用户现在只要花不到1000美元购买硬件和软件，就能够重现20世纪90年代录音棚的效果。

第五个主题是计算机处理器硬件的进步如何使数字处理应用成为可能，直到几年前，这些应用还被认为是计算领域之外的。摩尔定律表明，计算机处理器的性能每过十八个月就能提高一倍。这样做的结果是，在过去的十年里，处理器变得比以前强大了很多倍，再加上其他硬件的改进，比如数据存储设备速度的提高，这些意味着，诸如涉及飓风等自然世界的模拟，现在可以很容易地进行，而不需要配置强大的超级计算机。

第六个主题是计算机如何成为一种颠覆性技术，因为它既改变了许多技能，也消除了许多技能。这里的一个例子就是摄影。当我参观一个旅游景点时，我几乎从未看到有人使用胶片相机；手掌大小的小型数码相机几乎是现在的标配。此外，拍摄的照片可以放在记忆棒中带回家，放在家用电脑中，然后打印出来。相对便宜的程序，如奥多比公司的图片处理软件，现在可以通过调整曝光等方式来改善这些照片。

冲洗照片不再需要在暗房里把胶卷浸入化学溶液中。这显然是一种改进；然而，事情也有另一面，那就是摄影师的工作机

13

会减少了。有一个网站叫“网络相簿”(Flickr)。这是互联网用户上传照片并展示给访问者的照片共享网站。报纸编辑如果想为每期报纸购买廉价的库存照片(例如,为圣诞版购买一张知更鸟的照片),他们只需支付给自由摄影师一小笔费用。

第七个主题是不安全的计算机。一台没有连接到网络的独立计算机是完全安全的,不会受到任何技术攻击;计算机所有者应注意的唯一威胁是它可能被盗。然而,很少有计算机处于这种状态:大多数计算机都连接到互联网上。这意味着它们容易受到各种各样的攻击,从那些产生轻微公害效应的攻击,到可以完全停止计算机工作的严重攻击。这方面的一个例子就是僵尸计算机。它是指一台连接到互联网的计算机,已经被黑客、计算机病毒或特洛伊木马入侵。

僵尸计算机最常见的用途是充当邮件服务器,发送垃圾邮件;这类电子邮件试图向你出售一些你不需要的东西(比如万艾可、廉价股票或色情出版物),或者试图窃取你的银行账户身份等信息。大多数此类计算机的所有者,并不知道他们的系统正以这种方式被使用。这是因为主人往往并没有意识到他们被称为“僵尸”。2009年5月,安全公司“迈克菲”(McAfee)估计大约有1200万个新的僵尸计算机连接到互联网上。对于计算机入侵来说,这是一个相当惊人的数字。

14

一些例子

在深入研究这些主题之前,我们有必要先看看这些主题的一些实际例子。

挪威国家石油公司使用蓝色贻贝来监测石油钻井平台周围

的任何泄漏。当漏油发生时，贻贝的壳会收缩。考虑到石油钻探过程中泄漏对环境和收入的影响，挪威国家石油公司寻求一种方法来取代需要潜水器和深海潜水员参与的手工作业。他们所做的就是把射频识别（RFID）标签贴在蓝色贻贝的贝壳上。这是一种小型的硅基数据存储芯片，其中包含了一台计算机。当蓝色贻贝感觉到漏油时，它们就会收缩；这使得射频识别标签发出信号，表明该事件已经发生；这些信号被钻井平台上的计算机接收，然后停止导致泄漏的活动。例如，如果正在钻井，钻井线将自动停止。这种不同寻常的应用可能是计算机电路小型化进步的成果。

谷歌公司是广受欢迎的搜索引擎运营商。搜索引擎所做的事情之一是存储用户的查询，例如，您可以访问谷歌网站，并发现哪些是最受欢迎的查询。2008年，英国搜索量增长最快的是："iPlayer"、"facebook"、"iphone"、"youtube"、"yahoo mail"、"large hadron collider"、"Obama"和"friv"。这些术语大多与大受欢迎的网站或像iPhone这样的电子设备有关。最后一个词条"friv"是一个在线游戏网站。

15

稍后你将看到，从用户提交给搜索引擎的查询中可以获得大量信息。现在，警方调查人员通常会调查谋杀嫌疑人使用搜索引擎的情况。在谋杀案中，受害者的脖子如果被打断，他们就会检查诸如"脖子"、"绷断"、"折断"、"尸僵"和"尸体腐烂"等搜索词，这些词是凶手可能提交给搜索引擎的。

谷歌保存的用于查询的海量存储数据的一个有趣应用是跟踪流感病毒。两位谷歌工程师跟踪了诸如"温度计"、"流感症状"、"肌肉疼痛"和"胸部充血"等问题的发生率，并将查询这些问题的互联网用户的位置与美国疾病控制中心的数据库进行比较，发现了非常密切的相关性：实际上，他们发现涉及搜索词的查询量与流感病例的密度相似。你现在可以访问谷歌公司管理的一个网站，该网站显示了一段时间以来一些国家流感病例的增长。

这是本书关键主题的一个例子：计算机不仅可以访问自己硬盘驱动器上的数据，还可以访问存储在互联网计算机上的大量数据。

计算机应用的另一个例子超越了20世纪70年代和80年代的局限，涉及将计算机连接在一起以便协同工作的方式。 16

过去的二十年里，应用科学领域的研究人员一直试图从他们的计算机中获取处理能力。例如，人类基因组计划已经绘制出人类的基因结构，研究人员现在正利用这些信息来检测多种疾病的遗传原因。这项工作需要使用包含大量处理器的昂贵的超级计算机。然而，这一领域以及气候学等其他领域的一些研究人员则提出了一个新颖的想法，即邀请公众运行处理器密集型程序。

一个很好的例子是Folding@home。这个项目是观察蛋白质的结构，以检测治疗方案，可用于治疗阿尔茨海默病等疾病的患者。参与这个项目的研究人员已经招募了大约三万台家用电脑来分担计算负荷。志愿者们使用他们的备用处理器和存储容量来获取一个计算机程序的一小部分，让该程序得以进行蛋白质模拟，并将产生的结果反馈给协调处理过程的主计算机。

这并不是所谓的“大规模计算”或“大规模协作”技术的唯一应用。有些项目试图分析来自外太空的无线电波，以发现在我们的宇宙之外是否有智能生命，有些项目模拟原子和亚原子过程，还有许多项目与分子生物学有关。在过去，包含大量处理器的超级计算机仅供少量研究机构使用——实际上仍在被使用——然而，硬件进步和日益普及的宽带互联网意味着，我们都可以参与重大研究项目，而且对我们的家用计算机几乎没有影响。

“Wordia”是一个任何人都可以通过家用计算机访问的在线视觉词典。它当然包含单词，但每个单词都伴有一段视频，有人会告诉你这个单词对他们来说意味着什么。这是我遇到的最令人愉快的网站之一，也是所谓的大规模合作现象的一个例子。这是一个应用的例子，与本书的一个关键主题有关：计算机是松散耦合的全球计算机的一部分。

另一个例子涉及计算机电路的构造。当工程师们试图把越来越多的电子元件挤到硅晶片上时，这种芯片的设计就变得越来越困难；例如，把两个金属连接点放在一起会引起电气干扰，从而导致电路故障。考虑到数以百万计这样的电路可能被制造并嵌入到计算机中，一个错误对制造商来说将是非常昂贵的。设计的复杂性是如此之高，以至开发硅基电路结构的唯一可行方法就是使用计算机本身。

用来设计计算机电路的程序试图优化某些设计参数；例如，有一类程序试图挤压硅晶片上的连接，其方式是按照以下几个约束条件存储最大连接数：连接之间不要太近，电路的散热不会超过某个阈值，这会影响电路的可靠性。这里有许多用于优化的技术；最近有一类非常有效的程序是基于动物和昆虫的行为。

这方面的一个例子是一种被称作“群优化”的技术，在这种技术中，多个计算机进程相互协作，利用描述鱼群或鸟群行为的简单数学来发现问题的最佳解决方案。这是本书另一个主题的一个例子：程序员的聪明才智与大幅提升的速度相结合，使复杂的任务得以完成，这些任务在几年前甚至是无法想象的。 18

群优化是过去二十年中计算机应用革命的一个例子：它是由计算机的发展所代表的，从只是执行普通的处理步骤，如计算工资单的应用，到诸如设计计算机和控制内在不稳定的战斗机——这些已经成为我们军队的规范。

到目前为止，我主要关注的是计算机及其可见的用途。还有很多应用是看不到计算机的。离我最近的城市是米尔顿凯恩斯。当我开车去城市，然后绕着它非常实用的道路系统，我见识到了很多看不见的计算机应用。我经过一个由微型处理器控制的测速摄像机；一家利用计算机控制的机器人制造电子设备的公司；街灯由一台非常小的、原始的计算机控制；米尔顿凯恩斯医院使用的大多数监控设备没有嵌入式计算机就无法工作；还有购物中心，在那里，用计算机来保持每个商店环境的严格控制。

越来越多的计算机被用于隐藏的应用，在这些应用中，无论是硬件故障还是软件故障都可能是灾难性的，而且确实是灾难性的。例如，25号医用直线加速器是一台基于计算机的放射治疗机，它有许多软件问题。在20世纪80年代末，由于计算机接口的问题，许多病人接受了大量的过度辐射。

一个隐藏应用的例子，其失败可能是灾难性的，它是另一个有关控制石油钻机主题的例子。一个运转中的石油钻井平台将 19
极度易燃的石油或天然气从地下抽出，烧掉其中一些，并从石油中提取出无用的副产品，如硫化氢气体。在海洋设备中，这需要大量的人工操作人员。然而，越来越多的计算机被用于执行诸如控制石油或天然气的流动，检查是否有泄漏，以及调节燃烧过程等任务。

例如，曾经有信息技术人员入侵用于监控石油钻井平台操作的软件系统，或者是为了经济利益，或者是因为他们变得不满。人们没有认识到的是，尽管诸如传播病毒之类的计算机犯罪仍然普遍存在，但还有许多计算机应用也同样容易受到攻击。挪威智库辛泰夫集团报告称，海上石油钻井平台非常容易受到黑客攻击，因为它们转向了劳动密集程度较低、由计算机控制的操作——例如，用于远程监控钻机操作和通过卫星导航技术保持钻机位置的无线连接尤其脆弱。

本　书

本书的每一章都围绕着我在本章中概述的一个主题展开。

“小型计算机”将描述计算机体系结构是如何映射到硅晶片上的，以及计算机设计师在将越来越多的电子元件放到一块硅晶片上时必须面对的问题。这里和其他章节将要讨论的主题包括：超大规模的集成、硅制造、硬件设计过程及设计的新技巧和科技，比如使用人工智能程序最大化或最小化散热等因素。

20

“无处不在的计算机”将描述微型化如何使得计算机嵌入各种电子和机械设备。本章和其他章节讨论的例子包括：射频识别标签、超市会员卡的使用、用于监控年老体弱者的计算机、可穿戴计算机、虚拟现实系统中的计算机，以及手机、MP3播放器和计算机之间的融合。

“全球计算机”将着眼于互联网是如何使大量的计算机连接在一起，从而用来解决棘手问题，也就是那些在计算上很难解决的问题。这一章首先考察基因组测序的一个特殊的计算应用。然后，我将描述一个被称为网格计算的概念，在这个概念中，大量的计算机以某种方式连接在一起，以便它们的闲置能力可以用来解决这些困难的问题。本章的结尾将展望第七章，并简要描述网格概念是如何商业化到云计算的。这涉及将互联网视为一台拥有几乎无限计算能力和数据存储设施的巨大计算机。

“不安全的计算机”着眼于一些可能导致重大安全问题的威胁，包括技术和人的威胁。本章将涵盖攻击的完整图景，包括病毒攻击、木马攻击、拒绝服务攻击、欺骗攻击，以及那些由人为错误引起的攻击。本章将介绍可以使用的防御，包括防火墙、入侵探测器、病毒检查器和安全标准的使用。我的一个观点是，技术防御是不够的，它必须与传统的安全控制相结合。

21

“破坏性计算机”描述了计算机是如何产生重大破坏性影响的。大多数例子都描述了计算机和互联网中使用的通信技术相结合所造成的破坏。它将考察媒体行业，例如报纸，在过去的五年里是如何衰退的，以及网络广告是如何蚕食电视公司收入的。本章的结语部分将考察一些领域，在这些领域，计算机已经在去技能化，转化、改变或消除了某些工作。

“云计算机”描述了互联网是如何使开发人员和中等技能个人把这个网络当作一台大型计算机来对待的。许多公司（如亚马逊）都提供了对巨大产品数据库和编程设施的公共访问渠道，这样就可以开发跨多个领域的应用程序。这就引出了云计算机的概念：通过互联网连接的大量处理器和数据库，其软件接口任何人都可以使用。这一章介绍了软件混搭的思想：在这个过程中，复杂的应用程序可以通过集成或“混搭”组块既有软件来构建。

“下一代计算机”是一个相对较短的章节。它着眼于研究人员正在进行的一些“蓝天工程”，试图克服硅的局限性。它将专注于量子计算和生物计算。量子计算机是一种利用量子效应（如量子纠缠）对数据进行操作的计算机。虽然现在还为时尚早，但理论研究和一些早期实验已经表明，量子计算机有可能大幅提高处理速度。

这种计算机的影响可能是毁灭性的。举例来说，许多商业22计算都依靠密码科技，而这些技术都有赖于某些经典数字处理算法的巨大计算复杂度。量子计算机也许能够使这些算法变得可解，从而使互联网易受到攻击。

这一章还将描述DNA计算机背后的原理。这是当前计算机技术和量子计算机之间的一个中转站。DNA计算机利用生物链的遗传特性来提供非常大的并行处理工具。DNA计算机有效地实现了大量的硬件处理器，这些硬件处理器通过相互协作来解决困难的计算问题。

我希望在第四章和第七章中向你们传达的一个主要观点是，把计算机仅仅看作放在你桌子上的盒子，或者是嵌入诸如微波之类的设备中的一块硅，这只是一个片面的观点。互联网——或者更确切地说，宽带接入互联网——创造了一台巨大的计算机，可以无限制地使用计算机的能力和存储空间，甚至我们都认为永远不会从个人电脑迁移出去的各种应用也在这样做。

这方面的一个例子就是办公功能的迁移，如文字处理和电子表格处理——这是许多家用电脑的主要功能。谷歌公司已经推出了一套名为谷歌应用的办公工具。这些工具与微软办公软件中的工具类似，包括文字处理程序、电子表格处理程序以及类似于幻灯片的演示包等。

23

第二章

小型计算机

引　言

在过去的三十年里，计算机在处理速度、存储容量、成本和物理尺寸方面都有了惊人的进步。处理器的能力已经从20世纪70年代早期的90 kIPS提高到21世纪第二个十年的数千MIPS。处理器的速度用每秒指令数（IPS）表示，每秒指令数是计算机执行的某个动作，例如将两个数字相加；前缀“k”代表一千，而前缀“M”代表一百万。

存储容量也有所增加：我在前一章中描述过的埃利奥特803型计算机的存储容量为128千字节，放在一个有十二个棺材大小的柜子里：我的iPod存储容量为16千兆字节。

这种速度和存储容量的增长是如何发生的？在本章，我将回答这个问题；然而，在这样做之前，有必要简要地看看数据和

24 计算机程序是如何保存在计算机中的。

二进制数制

我们都习惯用十进制记数系统。当你看到一个数字，如69 126，它所代表的就是这个数字

$$6 \times 10^4 + 9 \times 10^3 + 1 \times 10^2 + 2 \times 10^1 + 6 \times 10^0$$

其中，每个数字表示自身乘以10的幂的结果（任何1次幂的数字，例如10^1，总是其本身，在本例中是10，任何0次幂的数字总是1）。

我们说一个十进制数的基数是10；这意味着我们可以用0到9之间的数字来表示任何十进制数。对于二进制数，基数是2；这意味着我们可以将二进制数如11011解释为

$$1 \times 2^4 + 1 \times 2^3 + 0 \times 2^2 + 1 \times 2^1 + 1 \times 2^0$$

它的十进制值是27（16 + 8 + 0 + 2 + 1）

数字就是这样以二进制形式存储的。文本也以这种形式存储，因为文本中的每个字符都有一个内部等效的数字。例如，美国信息交换标准代码（ASCII）是一种用于在整个计算过程中指定字符的标准。该代码为每个可以被计算机存储或处理的字符分配一个数字值——例如，A字符用等效于十进制数字65的二进制模式表示。

二进制系统也被用来表示程序。例如，

1001001110110110

可能表示将两个数字相加并将它们放在某个存储位置的指令。 25

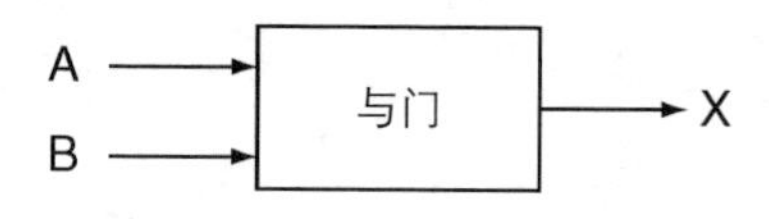

图2　与门的示意图

计算机硬件

计算机由若干个电子电路组成。最重要的是处理器：它执行包含在计算机程序中的指令。正如你在前一章中所记得的，有两种类型的存储器：用于存储相对少量数据的主存储器和用于存储大量数据（如字处理器文件）的文件存储器。

在计算机中还会有许多其他的电子电路：例如，如果你看一个台式机的背面，你会经常看到一个连接在金属条上的黑杆。杆子里有天线，可以从连接到互联网的调制解调器上接收无线信号。在金属条的后面是一个电路，它可以将天线接收到的信号转换成计算机可以使用的形式，例如，它可以在你的显示器上显示一个网页。

计算机由许多独立的电路元件组成。成千上万个这样的元件组合在一起，就构成了计算机处理器和其他电路。一个基本元素被称为与门，如图2所示。这是一个电路，有两个二进制输入A和B和一个二进制输出X。如果两个输入都是1，则输出为1，否则输出为0。它以一个表格形式来展示，这样的表格被称为真值表；图2与门的真值表如表1所示。

计算机内部有许多不同的电路——与门只是一个例子——当需要做一些动作时，例如将两个数字相加，它们就会相互作用来完成这个动作。在加法的情况下，对两个二进制数逐位处理

表1　与门的真值表

A	B	X
0	0	0
0	1	0
1	0	0
1	1	1

来进行加法。

那么，计算机是如何工作的呢？描述这一点的最佳方式，就是概述当我使用文字处理器时发生了什么。当我点击桌面上的MS Word图标时，视窗操作系统会感应到点击，然后将MS Word文字处理器加载到电脑的主存储器中。

然后程序开始执行。每当我执行一些动作时，文字处理程序就会感知到它，并执行它的部分程序代码。执行在所谓的读取-执行循环中进行。在这里，处理器读取每条编程指令，并执行指令告诉它做的事情。例如，一条指令可能会告诉计算机把我输入的内容存储到一个文件中，它可能会把一些文本插入到经过文字处理的文档的某些部分中，或者它可能退出文字处理程序。

无论程序（如字处理器）执行什么操作，循环都是一样的；一条指令被读入处理器，处理器解码该指令，对其进行操作，然后引入下一条指令。

因此，计算机的核心是一系列的电路和存储元件，它们可以读取和执行指令，并存储数据和程序。在过去的七十年里，各种技术被用于建造计算机。第一批计算机是基于电气继电器的。

27

这些都是机械开关，它有两种状态：二进制1由继电器关闭表示，而0由继电器打开表示。当你听到一个计算机程序员谈论他们程序中的一个“bug”时，这个术语就来自中继计算机的使用。1946年，计算机编程的先驱之一格雷丝·霍珀加入了哈佛大学的计算实验室，在那里她从事早期的中继计算机的研究。她描述了如何由一个程序错误追踪到一只被困在继电器中的飞蛾，从而产生了描述软件错误的“bug”一词。

第一代真正的计算机使用了基于阀门设备的电子电路。这些看起来有点像小灯泡，它们可以接收一个发送到阀门的信号，通过电子方式从0（关）状态切换到1（开）状态。程序员使用纸带或穿孔卡与这些计算机进行通信，这些卡保存了待处理的数据或正在进行处理的程序。

在早期计算机中的主存储器使用了被称为磁芯的圆形磁性材料。这些存储的二进制1或二进制0取决于它们的磁化状态。

第一代计算机被使用晶体管的第二代计算机所取代。晶体管是一块可以打开和关闭的硅；我在前一章中描述的埃利奥特计算机就依赖于这种技术。

从20世纪60年代末至70年代初，物理学家、材料科学家和电子工程师设法将第二代计算机中由晶体管实现的电路沉积到硅晶片上。这个过程在小型计算机中被称为超大规模集成电路（VLSI）。这些第三代计算机就是我们今天使用的计算机。正是超大规模集成电路技术为今天的计算机提供了令人难以置信的小型化、速度和容量。以元件间宽度为代表的小型化，已经从20世纪90年代初的1.00微米左右下降到21世纪初的40纳米。一微米相当于一米的百万分之一，而一纳米相当于一微米的千

28

分之一。

计算机电路

现代计算机硬件依赖于硅。要将一块硅转换成处理器或接口电路（如用于驱动计算机显示器的接口电路），需要经过许多制造步骤。

第一步是单晶硅生长成为晶棒。当晶棒生长完成后，被称为“晶圆”的圆形薄片就可以从晶棒上切下来，就像你从一筒午餐肉上切薄片一样，唯一的区别是薄片通常比肉片薄一些。切片切好后，再进行抛光。

下一步是在每个硅晶片上嵌入要实现的电路的设计。这是通过一种被称为掩模版的设备来实现的，其将在晶圆上展示电路图案和电路元件的网格。紫外光通过栅极照射到晶圆的一部分上，这就形成了要沉积在晶圆上的电路的指南。通常，在硅晶片上蚀刻许多类似的电路图。

具体的制作过程如下。首先，硅晶片在充满氧气的烤箱中烘烤。这在表面形成了一层薄薄的二氧化硅。然后，再在晶圆上涂上一层被称为抗蚀剂的薄有机材料。于是，我们现在有一个硅基，通常被称为衬底，一个二氧化硅层，以及顶部的抗蚀层。

然后，紫外光通过掩模版照射到晶圆的一部分上。抗蚀层的结构被光破坏，但其余层不受影响。这个过程非常类似于黑白照片的冲洗方法。一旦光照射到晶圆的一部分上，掩模版就会移动，晶圆的下一部分就会被蚀刻图案。

下一阶段是晶圆的制造。这需要把它和许多其他晶圆放在溶剂中，溶剂会溶解已接收到紫外光的抗蚀层部分。

晶圆现在有一层硅，一层二氧化硅，以及抗蚀层中不受紫外光影响的部分。被溶剂除去的那部分晶圆将暴露出二氧化硅的部分。然后，通过使用另一种溶剂将这些移除，以使底层的硅暴露出来。

现在硅晶片将包含一层暴露的硅部件；二氧化硅层，它将有部分被切掉，以暴露出硅；以及与二氧化硅相同的部分被切割的抗蚀层。

下一步是使用一种溶剂来溶解它以去除抗蚀层。晶圆现在包含一个基层的硅与蚀刻的电路图案。然后，对硅层的暴露部分进行处理，以使其能够传导电信号。晶圆现在已经处理了代30 表电路的硅和作为小型绝缘体的新二氧化硅，以确保通过硅的一部分信号不会影响其他信号路径。

然后再沉积一层，以完成电路的最后一层二氧化硅层。这上面蚀刻了孔，以便能与基础电路连接。

封装电路的过程现在开始了。封装技术有很多种。我将描述最简单的。首先，在每条电路的边缘沉积方形金属接头，称为衬垫。然后将另一层二氧化硅放置在晶圆上，在该层上蚀刻小孔，以便与焊盘连接。

然后，每个电路由一个特殊用途的电子设备进行测试，该设备将与焊盘接合，并向一些焊盘发送信号，监测信号对其他焊盘的影响。任何测试失败的电路都会被标记上染料，最终被拒绝。如果电路通过测试，另一层二氧化硅或氮化硅被放置在电路上，并在这一层与衬垫形成连接孔。这最后一层起到了物理保护的作用。

最后一步是从硅晶片上切割出每个电路。这是通过机械刀具实现的；这在概念上类似于玻璃匠将玻璃切割出一种形状的方式。晶圆现在已经变成了一组相同的芯片。

最后一步是将每个芯片安装在某种框架中，以便能够装入计算机。有各种各样的技术可以做到这一点。一个简单的方法是使用一种黏合剂将芯片粘在引线框架上，这种黏合剂有助于将芯片上的热量传导出去，然后将信号线放在芯片上，与芯片上的衬垫连接。在电线添加进去之后，芯片就会被一些塑料材料覆盖，作为最后的保护。

如果你对更多的细节感兴趣，那么克莱夫·马克斯菲尔德所著的《电子学非传统指南》是一本很好的介绍计算机电子学的书。

计算机存储器

有两种类型的存储设备：只读存储（ROM）设备和读写存储（RWM）设备。前者保存不能更改的数据；后者可以擦除和重写数据。

计算机存储器是以硅的形式实现的，其制造方法与硬件处理器和其他电路的制造方法相同。计算机存储器与互联网或处理器通信电路之间的唯一区别是，前者有一个规则的结构。

图3显示了典型的存储器布局。它由一组晶体管组成。每个单元格可以包含一个0或一个1。每个水平的单元格集合被称为一个词，数组的深度则是指数组的嵌套层级。

一个称为地址总线的电路被连接到阵列上。数组中的每个词都有一个唯一的标识，即它的地址。当计算机的处理器需要来自存储器的数据或程序指令时，信号沿着总线发送；它指示存

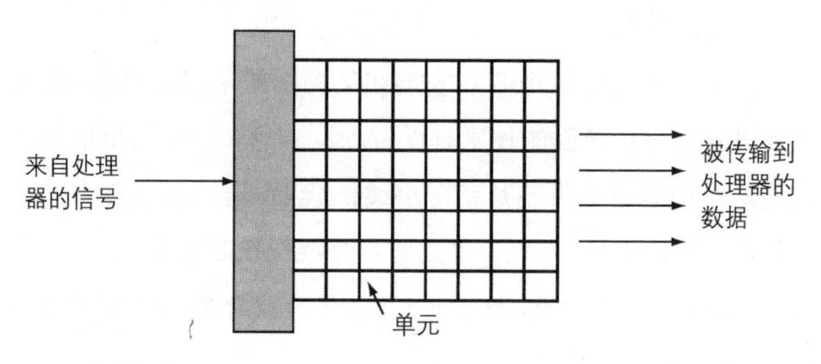

图3 计算机存储器

储单元对地址进行解码，并使数据或程序指令在处理器指定的位置可用。

可用的存储设备有很多种。掩码编程的ROM在制造时将其数据或程序放入其中，不能更改。可编程只读存储器，通常称为32 PROM，它们是空白的，然后可以用电子设备编程。然而，由于它们是只读的，所以这只能发生一次——它们不能被重新编程。

可擦可编程只读存储器，通常称为EPROM，比PROM更进一步，它可以擦除然后重新编程。关于EPROM存在着一些困惑：由于它们可以被重新编程，所以人们认为，当它们在计算机中时，它们的内容是可以通过覆盖来改变的。覆盖只能通过已插入EPROM的专用设备来实现。

电可擦可编程只读存储器（简称EEPROM）的发展，是连接只读存储器和可写存储器世界的重要一步。这是EPROM的一种形式，当它成为计算机的一部分时可以被擦除。利用与EEPROM相关的思想和技术，一种被称为闪存的技术得以开发出来。

图4　硬盘单元

闪存技术在需要大量不可擦除存储器的应用中被采用。例如，它被用于USB记忆棒，用来将数据从一台计算机传输到另一台计算机，或者作为备份存储，以防计算机崩溃。其他使用闪存技术的应用包括个人数字助理、笔记本电脑、数字音频播放器、数码相机和移动电话。被称为上网本的新一代小型笔记本电脑，就使用闪存存储程序和数据文件。

可写存储器是使用许多技术来实现的。动态随机存取存储器（DRAM）是计算机中最常用的存储器。它的实现方式可能会在短时间内丢失数据。因此，DRAM单元的内容被不断地读取和写入，以便恢复其数据。

静态随机存取存储器（SRAM）是这样一种技术，它不需要刷新以应用到它的单元，除非一个程序改变了单元，或者电源从安装它的计算机中被移除。它比动态随机存取存储器更快，但

34 也更为昂贵。

文件存储技术

到目前为止，我所描述的技术通常用于相对较小的数据量（这是微型化和制造技术进步的显著结果，我可以将8千兆字节称为“相对较小”）。对于大量的数据和程序，则使用一种不同的、速度慢得多的技术。这就是所谓的硬盘技术。

本质上，硬盘单元由一个或多个可磁化的圆形金属盘组成。每个磁盘有大量的可磁化区域，根据磁化程度可以表示0或1。圆盘在高速旋转。该单元还包含一个或多个臂，可以横向移动，可以感知磁盘上的磁模式。硬盘单元的内部如图4所示；在这里，可以清楚地看到靠近圆盘边缘的臂。

当处理器需要一些存储在硬盘上的数据时，比如一个字处理器文件，然后它发出一个指令来查找该文件。操作系统——控制计算机的软件——将知道文件的起始和结束位置，并将向硬盘发送消息来读取数据。该臂将横向移动，直到文件的起始位置之上为止，当旋转磁盘在该臂下通过时，表示文件中保存的数据的磁模式被它读取。

访问硬盘上的数据是一个机械过程，通常只需要几毫秒的时间。与计算机本身的电子速度（通常以微秒为单位测量）相比，这个速度慢得令人难以置信。

由于磁盘访问速度较慢，系统设计人员试图尽量减少文件35 所需的访问量。一种特别有效的技术被称为缓存。例如，它可以用于网络服务器。这些服务器存储发送到浏览器以显示的网页。如果你观察与一个网站相关的网页访问模式，你会发现有些网页被检索得非常频繁——例如主页——而有些网页被访问得很少。缓存包括将频繁访问的网页放置在一些快速存储介质中，如闪存，并将其余网页保存在硬盘上。

克服硬盘访问速度慢的另一个方法是，用闪存等电子存储介质完全取代硬盘。目前，这类技术的存储容量和成本还没有接近硬盘单元：通常闪存可以买到64千兆字节，同样的价格你可以买到500千兆字节的硬盘。然而，对于某些类型的计算机，其存储需求适中，电子存储器用于大量文件存储现在是可行的。例如，一些计算机的特点是电子存储介质的大容量存储，而不是硬盘。这种计算机包含低功耗的处理器，而这种存储器的使用补偿了这一点。

未来的技术

在过去的三十年里，小型化取得了惊人的进展。然而，有些性能和规模将很快达到瓶颈。例如，当你把越来越多的组件封装到一个硅晶片上时，它们变得更容易出错，因为产生了随机信号；另一个问题是，一种被称为亚原子侵蚀的现象开始出现，破坏了硅的结构。高度小型化的电路也会出现设计问题。

在最后一章，我将讨论两个激进的想法，如果成功的话，它们将改变计算。这包括在电路中使用生物材料，以及在处理器开发中运用量子物理思想；这需要非常长期的研究。然而，目前正在研究和开发的是用于计算机电路的其他低水平技术和材料。它们包括使用光刻技术来生产更快的组件，组件之间使用光学连接，用碳取代硅，以及在计算机电路中使用超导材料。

36

37

第三章

无处不在的计算机

四个别出心裁的例子

几年前，我把车提前停在了米尔顿凯恩斯，等着使用自动售票机时，我注意到前一位顾客是停车场服务员。我问他，如果他不得不把车停在米尔顿凯恩斯，他的工作成本是不是太高了，他是否得到了员工折扣。他告诉我，他拿停车票的原因不是把它贴在汽车的挡风玻璃上，而是作为他在指定时间开始工作的证据——他实际上是把这台机器当作一个时钟。

在《咖啡机里的间谍》一书中，奈杰尔·沙德博尔特和吉隆·奥哈拉描述了计算机在日本的一种创新应用。日本人有一个主要的问题是老龄化趋势：该国的出生率是世界上最低的国家之一，而且不鼓励移民。因此，日本的人口正在迅速老龄化，照顾老人的需求也在增加。实现这一目标的一种方式是通过基于计算机的设备。其中之一就是iPot。这是一种可以让咖啡或汤全天保温的水壶。每当使用iPot时，其就会向服务器发送一

条消息，每天两次，使用频率的报告通过手机短信或邮件发送给亲属或看护者，让他们知道那些被照顾的人都很健康。 38

日本电气公司开发了一种新型眼镜，可以将信息投射到用户的视网膜上。这种设备有多种用途，其中的两个例子是，帮助在呼叫中心使用计算机的员工，以及当用两种不同语言进行对话时提供实时翻译。（正如你将在第七章中看到的，计算机语言翻译已经发展到可以获得相当不错结果的地步，所以这并不是一个真正的科幻应用的例子。）

荷兰皇家德克兹瓦格公司开发了一个系统，可以实时跟踪全球船舶的航行。它使用一种基于卫星的技术来实时跟踪船舶的位置。它使得船舶可被引导到有空间的港口，减少了船舶等待泊位的时间，并减少了船舶所需的燃料量。

所有这些例子都来自计算机微型化和制造成本降低相结合的独创性，我在第二章中详细介绍了这一点。除非你是一个隐士，否则它们只是你在一天中遇到各种各样计算机的小快照。

在写这本书的时候，我在我的村庄周围散步了大约四十分钟：我发现这让我的头脑清醒，给了我思考的时间。其中一条是在我家附近一条僻静的乡间小路。8月的时候，我沿着这条小路散步，试着思考不接触计算机的情况；我开始认为这次散步是为数不多的例子之一。事实并非如此：当我走过树篱的缝隙时，我注意到远处有一台联合收割机在工作。这种收割机使用精密的计算机控制来调节收割机的前进速度，并保持脱粒滚筒的转速 39 不变：这是一件相当棘手的事情，电气工程部门的研究人员仍在努力开发技术，以优化脱粒效率。也许假使我住在南极？即使在那里，也很难摆脱计算机：科学家们使用计算机标签来追踪企鹅的活动。

有一点是明确的，无论我们走到哪里，我们都会与计算机接触，它们越来越多地改变了我们的生活。例如，加州大学圣迭戈分校的研究人员最近进行的一项研究估计，每天传递给我们的信息量大约相当于10万个单词的文本。在本章中，我将探讨这种信息超载如何影响科学家，探讨在医疗保健中计算机无处不在的一些积极方面，并指出一些与隐私和保密相关的问题。

计算机无处不在

有一些趋势已经把计算机从个人电脑中解放出来。首先是电子元件的日益小型化——不仅仅是硬件处理器和存储器，还有通信电路和用于信号监测的电路。其次是计算机之间无线通信技术的发展。第三，也是经常被忽视的一点，是电子电路的坚固性增加了：我的手机掉了很多次，但它仍然能用。

坚固性意味着计算机可以安装在几乎任何地方，即使在最极端的条件下也能正常工作；例如，气候变化研究人员已经在货船和油轮上安装了基于计算机的测量仪器，以便在它们航行的过程中测量海洋的温度——这些计算机会受到海浪的冲击，必须经历温度的重大变化，还会受到引擎振动的影响，但仍能正常工作。

计算机无处不在：iPod、移动电话、安全系统、汽车导航设备、自动取款机、汽车电子电路等。这有三层主要的含义。第一层含义是，它启发了一门新的环境信息学学科，在这门学科中，数据可以在任何时间和任何地点获得。

第二层含义是，因为这些数据是由我们在世界上正常的交互产生的，例如通过访问商店感官计算机设备，我们可能会携带或将设备嵌入一副眼镜，或者通过使用智能卫星导航系统开车去一些目的地，这对隐私和安全有重大影响。

第三个含义是关于交互的方式。我以一种人为的方式与我的个人电脑互动，在这种方式中，我总是意识到有一种互动形式正在进行。普适计算所涉及的交互在某种意义上是自然的，因为它不引人注目。这里有一个例子。普适计算的第一个应用是与物理安全相关的，它涉及将计算机嵌入一个名为“活动标识”的身份标识中。计算机发出的信号被建筑物内的监测点接收，并向员工提供有关同事或访客所在位置的信息。佩戴一个电子识别卡是不引人注目的：你不会感觉到正在发射的无线信号。它还不如你偶尔意识到的心跳那么突兀。

为了研究其中的一些含义，我们有必要将重点放在一种成熟且廉价的环境技术上，这种技术有许多工作应用。 41

射频识别

射频识别技术，或者更为人们所熟知的简称RFID，涉及将一个小的——它们可以非常小（布里斯托尔大学的研究人员已经将射频识别设备连接到蚂蚁身上）——电子设备连接到物体或人身上。射频识别设备，即标签，发出的无线电波可以被一台装有无线电接收器的计算机接收。射频识别标签最初用于库存控制，零售商（如超市）将其附在库存商品上，以跟踪销售和库存水平。

这些标签现在零售价为几便士，它们的使用已经从库存控制领域转移。例如，射频识别标签在医院被用于追踪可能急需的设备；它们可以附在年幼的婴儿或患有痴呆症的患者，以便发现其当前的位置；它们可以用作医疗信息的存储库，即当病人进入医院时，给他们一个附在腕带上的标签，这样的话，当他们参加一些医疗检测时，检测数据可以立即下载到标签上；它们可以用来为田径等运动的参与者计时；它们代表着一种潜在的技术，可以应用于智能交通管理系统中，它们可以附在汽车上。

显然，射频识别标签有很大的潜力，但也有缺点，即隐私问题。由于产品上的射频识别标签在购买后仍可使用，而且可以用于监视和其他用途，因此曾有多起抵制射频识别产品的事件。加利福尼亚州参议员黛布拉·鲍恩在一次隐私听证会上表达了一种典型的担忧：

42

> 比如说，如果有一天你发现你的内裤在报告你的行踪，你会怎么想？

显然，附在食品包装上的射频识别标签不会带来什么隐私担忧：当食品被吃掉时，包装通常会被处理掉。然而，附在移动电话、MP3播放器等设备和其他物品（如衣服）上的标签，以及位于购物中心、收费公路和其他公共场所的无线阅读器，则为监控社会提供了技术基础。

然而，当一件物品被带到维修店时，这样的标签可能非常有用。有一个解决这一问题和隐私担忧的建议是，当物品被购买时，包含射频识别标签的一部分可以撕掉，而留下基本数据，只能由一个手持阅读器读取，并且只在几厘米之内有效。

此外，人们还对可植入人体的射频识别标签的潜在用途表达了隐私担忧。一家名为“绝对芯片”（Verichip）的公司开发了一种射频识别标签，可以通过手术植入人体。这种芯片得到了美国食品和药物管理局的批准。显然，这种类型的标签有一些应用，例如持续监测重要功能，如血压；然而，它的广泛使用也引发了有关隐私的问题。

射频识别技术代表了最先进的普适计算思想的有效运用。然而，还有许多其他的技术几乎同样先进，或者已经有所超越；其中一些值得一看。

健　康

21世纪的主要趋势之一是出现了越来越多的集成，计算机在硬件之间扮演着数据处理器和同步器的角色。这方面最好的例子就是iPhone，它集移动电话、个人记事本、MP3播放器和上网设备于一体。

在《纽约时报》（2009年11月5日）的一篇文章中，专栏作家戴维·博格描述了他如何被要求在泰德医学会议上就苹果手机的医疗应用程序做十八分钟的演讲。泰德会议的组织方是一家非营利性组织，其主要目的是传播思想（他们的网站非常棒，网址是http：//www.ted.com/，包含了大部分演讲的视频）。

博格本来担心他找不到足够多可以谈论的应用程序，从而无法填满那十八分钟。最终他遇到的问题是，他发现了太多的应用程序：超过7000个应用程序，而且只针对苹果手机。对于普通用户来说，它们包括：“Uhear”，一款能让苹果手机测试听力的应用程序；“ProLoQuo2Go”，一款语音合成器，可以让有语音障碍人士触摸苹果手机屏幕上的短语、图标和单词，然后说出他们识别出的文本；“Retina”，一款可以让色盲用户将苹果手机指向某个有色物体，然后识别出该物体颜色的应用程序。

博格还发现了供医疗人员使用的应用程序，其中包括：“Anatomy Lab”，一款针对医学生的虚拟尸体，允许用户探索人体解剖；“Epocrates”，一款可以提醒医生两种处方药可能产生副作用的电子百科全书；“AirStrip OB”，它能让产科医生远程监测孕妇的生命体征。

普适计算在老年人护理方面也有重要的应用。在一次关于辅助护理技术使用的国际会议上，四位来自得克萨斯大学计算机科学系的研究人员，描述了一种简单而廉价的无线网络的设计，这种网络可以应用在养老院或老年人家中。这与之前详细介绍的监控日本老年人在家状态的iPot类似。

44

他们所描述的网络可以支持各种各样的监控，从而轻松实现各种各样的应用程序。比如监测运动，当一段时间没有探测到运动后就会发送消息到远程监测站；又比如监测生命体征，使用射频识别技术监测心率。未来十年，消费电子领域的一个主要扩张领域将是基于无线的家庭娱乐，那时所有的有线连接将被无线连接取代。研究人员所描述的这种网络，可以很容易地附加在这些局域网上。

全球“圆形监狱”

计算机可以放置在各种各样的地方：卫星、远程望远镜、温度测量系统，甚至装载在货船上。在过去的十年里，地质学、气候学和自然地理学等领域的研究人员所能获得的数据量大幅增加。

有大量的例子表明，嵌入硬件的计算机可以产生大量的数据。澳大利亚“平方千米阵列”望远镜将产生数十亿的数据项，天文望远镜的“泛星计划”阵列将每天生成数拍字节的数据。基因测序技术也正在迅速推进，数十亿的脱氧核糖核酸（DNA）数据项可以在短短几天内生成。

这些计算机的输出具有巨大的价值，例如说改变了我们对气候变化的研究。然而，虽然计算机提供了这样的数据，但它也带来了一些重大问题。核子对撞机、基因测序仪和射电望远镜等设备提供的数据对整个科学界都有价值，而不仅仅是那些使用这种硬件进行实验的人。不幸的是，对于这种数据的存储和发布，几乎没有标准。

45

世界上只有为数不多的科学数据图书馆。加州大学圣迭戈分校的圣迭戈超级计算机中心就是一个很好的例子。这里存有包括生物信息学和水资源实验在内的大量数据（目前有很多拍字节），并以其他研究人员可以访问的方式存储在这里。另一个例子是澳大利亚国家数据服务局（ANDS），它提供数据注册服务。ANDS不存储数据，它存储关于其他地方存储数据的信息：可访问数据的网站、数据的性质和谁负责生成数据，这是可以通过ANDS计算机访问的三个项目。

不仅是数据量在增加，研究文献也在增加。医学研究就是信息量暴涨的一个很好例子。1970年，大约有20万篇研究论文被编目；到2009年，这一数字已上升至80万。

在撰写本章初稿的过程中，一个重大的安全事件影响了东安格利亚大学气候研究所（CRU）。CRU是世界上最重要的气候研究机构之一，负责多个数据库，其中一个包含来自世界各地温度测量站的数据。

2009年10月，黑客从CRU的一台计算机中窃取了大量电子邮件、文档和程序描述文件。这一盗窃事件发生在哥本哈根气候变化大会召开前几周，这次文件盗窃引起了博客帖子、文章和电子邮件的议论狂潮。在“气候门”事件的鼎盛时期，我发现有超过3000万条关于它的引用。

“气候门”事件有几个维度：关于CRU的科学家是否篡改了数据，CRU科学家的行为是否与其他不认同他们对全球变暖观点的科学家相违背，以及英国的《信息自由法》是否存在争议。

然而，从这一事件中可以清楚地看出，世界各地的气候数据具有不同的性质，各种数据存储库缺乏适当的编目，而且缺乏操纵这些数据的可见程序代码。

如果科学界想要在充分利用计算机的数据收集和数据处理方面取得任何进展，那么就需要做出重大的改变，就像政治的、智识的、科学的努力造就了多学科的团队，这些团队为曼哈顿计划或布莱切利密码破译项目而团结在一起，并对第二次世界大战的结果产生了重大影响。

在那些迫切要求将互联网用作科学数据开放存储库的人中，微软公司的研究员吉姆·格雷是最重要的。格雷是一名计算机科学研究人员，他强烈要求科学界接受计算机给他们的研究带来的变化。他设想了第四种范式，与现有的三种科学范式并列，即经验观察、数据分析处理（通常使用统计方法），以及基于分析的模拟或建模，由此创建和修正与生成数据现象相关的理论。格雷的第四种范式有许多组成部分，这些组成部分将会带来技术和政治上的挑战。

第一个组成部分是对实验数据的整理。这涉及专门用于存储数据、元数据和任何与数据相关的计算机程序的互联网站点的开发，例如，在数据存储到数据库之前对数据进行一些操作的程序。这是一个重大的挑战：它不仅涉及将大量的数据存储在一个网站，而且还涉及规范元数据——描述每个数据集合是什么意思的数据——例如，事实上一系列的数据来自两个日期之间一个特定的测温站。它还包括存储数据如何更改的细节，以及用于影响更改的流程和程序。

格雷第四范式的第二个组成部分与研究出版物的爆炸式增长有关。如果你看了一篇研究论文的结构，就会看到其他论文的参考文献、数据集的参考文献，以及计算机程序。阅读和消化这些论文所需要的脑力劳动越来越多。

研究人员正在着手解决这个问题。帮助读者的一种方法是在原始文章中添加支持材料的增强形式。欧洲分子生物学实验室（EMBL）德国的“映射”工具就是这样做的一个例子。它在一篇研究论文中标记基因、蛋白质和小分子的名称，这些标记与保存生物信息数据的相关序列、结构或相互作用数据库超链接。因此，如果读者想要交叉引用一个基因序列，他们所要做的就是点击论文中的链接。

格雷第四范式的第三个组成部分是数据可视化。当你有大量的数据时，在处理了这些数据之后，将有稍少但仍然大量的输出结果。在越来越多的情况下，这种输出的规模正在击败传统的显示方式，例如二维图。研究人员现在正努力开发新颖的——通常是三维的——渲染输出的方法，这样读者就可以辨别模式。 48

格雷第四范式的第四个组成部分是整合组成一个基于计

算机的科学实验的所有元素。当你做一个实验时，首先对实验将做什么有一个概念性的想法，然后把这个想法转变成一个具体的过程，包括收集数据等，然后用一些计算机程序处理数据，并显示结果，在最后一步，将研究结果发表在学术期刊或会议上。这需要一个翔实记录，它描述这个过程中的每一个步骤，以及链接到的所有相关文件、数据和实验中使用的程序代码。毫无疑
49 问，这是那些在实验中使用计算机的科学家所面临的最大挑战。

第四章

全球计算机

简　介

在本章中，我希望能使你相信，把计算机看作放在桌子上的盒子，或者是一块用于控制和监控应用（如航空电子设备应用和化工厂监控）的硅是有局限性的。我希望能让你相信，通过连接计算机——它们的处理器和存储器——我们实际上可以创造更大的计算机；这一现象的最终实例就是互联网。

全球计算机发展的关键是它的处理器：一种电子电路，它读取、解码和执行计算机程序中的指令，并实现程序员的意图。在过去的五十年里，计算机的速度增长了几个数量级。一旦计算机技术在性能方面（处理器速度、存储器大小和内存访问速度）有所进步，新的应用程序也会随之发展，它们需要更快的速度、更大的存储，或更快地访问内存，或者是对当前应用程序有了改善需求，如天气预报，硬件的进步使预测更加准确，并使预测者能够深入到未来。　50

棘手问题

在研究我们如何创造出越来越强大的计算机之前，有必要先看看它们必须解决的一些主要问题——所谓的“棘手问题”，这些问题的解决需要大量的计算资源。这个世界充满了棘手问题，它们需要大量的计算机资源和人力才能解决。其中之一就是人类基因组计划。这个项目发现了基因序列。如果获得的序列要存储在书中，那么大约需要3300本大书来存储完整的信息。搜索基因数据库，以寻找使人容易患上某种疾病或状况的基因，这需要的计算资源是巨大的，并需要超级计算机。

后续的项目进展缓慢，因为计算需求巨大，只能由我在本章后面描述的超级计算机来满足，但它仍在进展中。然而，有一类问题是不能完全由计算机来解决的，它们被称为“NP难题”。

这些问题的一个令人惊讶的特点是，它们很容易描述。这里有个例子叫作集合划分问题。它涉及判断是否可以将一组数字划分为两个子集，使每个子集的数字之和相等。例如，集合

（1，3，13，8，6，11，4，17，12，9）

可以划分成两个集合

（13，8，9，12）

和

51

（4，17，6，1，3，11）

每个加起来是42。这看起来很简单，对于小集合来说确实

如此。

然而，对于更大的集合，例如包含数万个数字的集合，发现这些集合是否可以拆分所需的执行处理时间是令人望而却步的；它会很快就达到这样一个点，即使是最强大的计算机已经被制造出来，所需时间也将超过已知的宇宙生命。

这类问题不是学术性的，它们常常产生于实际应用；例如，NP难题出现在超大规模集成电路设计、遗传序列分析和航空电子设备设计中。最著名的NP难题之一是旅行推销员问题，它产生于一个与计算机硬件设计相关的任务。这里的目标是，给定一系列城市和它们之间的距离，为某人（旅行推销员）开发一条路线，让他们在同一时间到达每座城市，同时最小化路线，从而减少汽油使用量。

在实践中，绝大多数NP难题不需要一个精确的解决方案——一个接近精确解决方案的解决方案就可以了。例如，有一个NP难题，称为装箱问题，在这个问题中，给计算机一些容器和一系列盒子，目标是使容器中的闲置空间最小。对于这个问题，有可能得到99.5%的最优解决方案。

正因为如此，大量与NP难题相关的研究关注的是所谓的近似算法。顾名思义，这些是对计算机程序的描述，这些程序可以产生近似但足够好的解。

52

用软件解决棘手问题

最近，软件研究人员利用生物学中的一些想法来提高程序的能力，这些程序试图产生近似的解决方案。其中之一就是遗传编程。这里生成一组候选程序来解决问题，然后运行。效率最高的程序被收集起来作为新一代的程序，并结合在一起创造出更新一代的程序。这一代程序接着运行，并做进一步的接合，直到一个有效工作的合适程序出现。

“遗传编程”一词源于这样一个事实，即生成越来越有效程序的过程模仿了达尔文的进化过程（遗传编程通常被称为进化编程）。与我在本章中详细介绍的其他技术一样，它需要大量的计算机资源。

为了解决棘手问题，还有许多其他技术可以模仿生物过程。群计算是基于一种模型，该模型利用了鸟在群体中所表现出的行为，昆虫在一项任务（比如觅食）中相互合作，或者鱼在鱼群中游动。这些行为可以用非常简单的方式建模，而这种简单性已经被转移到许多优化程序中。

还有一些计算机程序可以模仿蚂蚁等群居昆虫的行为，例如，它们处理数据的方式与蚂蚁寻找食物或处理尸体的方式相同。这种蚁群程序解决的问题与一个被称为路由的区域有关；在这里，底层数据可以被模型化为一系列由线连接的点，例如在超大规模集成电路制造芯片上的连接布局。

还有其他方法可以解决一些棘手问题。计算机科学中有一53个备受争议的领域被称为人工智能，因为它的支持者在过去的三十年里所做的过度声明。这一领域的研究人员分属于两个阵营：一类阵营试图利用计算机来理解人类如何进行推理等过程的，另一类阵营只想开发出性能与人类相匹配的智能软件人工制品——不管软件是否与人类过程有任何相似之处。

人工制品构建者已经开发了许多在现实世界中获得成功的技术。几乎可以肯定的是，这种人工智能方法最著名的产品是

图5　克雷巨型机XM-P48

54

“深蓝”，这是一个国际象棋程序，它在1997年击败了世界象棋冠军加里·卡斯帕罗夫。

这个程序依赖于一台超级计算机的巨大计算能力，但并没有把它的大部分能力建立在研究人类国际象棋玩家行为的基础上。然而，有许多人工智能程序试图通过蛮力计算和人类启发式的结合来战胜棘手问题。程序的一种类型被称为专家系统。它试图执行人类执行的任务，如诊断疾病或发现电子系统的硬件故障。这种专家系统不仅依赖于计算机的性能，而且依赖于人类专家在其工作领域中使用的一些启发式编码。

即使考虑到软件技术的进步，我们还是见证了过去的二十年仍然需要强大的计算机：软件技术在解决大问题上已经走过一段距离——但还不够远——并且有自然趋势继续前进和奋力解决更大问题、更复杂问题。确实如此，过去的四十年见证了超级计算领域的重大进步。

超级计算机

第一批计算机只有一个硬件处理器，可以执行单独的指令。不久之后，研究人员开始考虑拥有多个处理器的计算机。这里的简单理论是，如果一台计算机有n个处理器，那么它将会快n倍。55在讨论超级计算机这个话题之前，有必要先澄清一下这个概念。

如果你看看很多你认为可以用超级计算机来解决的问题，就会发现严格的线性性能增长并没有实现。如果一个问题可以由单处理器计算机在二十分钟内解决，那么你会发现双处理器计算机可以在十一分钟内解决它。一台三处理器计算机可以在九分钟内解决它，一台四处理器计算机可以在八分钟内解决它。有一条收益递减的规律；通常，添加处理器会降低计算速度。实际情况是，每个处理器需要与其他处理器进行通信，例如传递一个计算结果；当添加的处理器超过所做的有用工作数量时，这种通信开销就会越来越大。

有效的问题类型是，一个问题可以被分解成子问题，这些子问题几乎可以由每个处理器独立解决，几乎不需要通信。

超级计算的历史可以分为两个阶段：20世纪70年代和80年代，以及这二十年之后的几年。在回顾历史之前，看看过去七十年来超级计算机的速度是如何提高的将很有启发意义。

第一批真正的超级计算机是由美国数据控制公司和克雷公司开发的。最成功的设计是基于矢量架构。这是基于一种能够同时对数据执行许多指令的处理器，例如同时将1000个数字相加。克雷公司制造的计算机是标志性的超级计算机。图5显示了克雷巨型机，其中之一就放置于欧洲核子研究组织（CERN）。它类似于为20世纪80年代一家广告公司的接待室设计的家具。然而，当它交付给世界各地的研究实验室时，它是当时最快的计算机：1982年，它是一台最先进的计算机，两个处理器的理论最高速度为800 MFLOPS（一个MFLOP是每秒执行一些算术运算的一百万条指令，例如将两个数字相加）。

超级计算机被送达各种各样的客户，包括欧洲核子研究组织、美国洛斯阿拉莫斯国家实验室、波音公司、英国气象办公室、日本国家航天实验室、美国国家核安全管理局和美国能源部橡树岭实验室。

这些庞然大物的客户泄露了它们的用途：核实验模拟、天气预报、模拟核反应堆的日常运行过程，以及大型飞机的空气动力学设计。这些应用之间的关键相似之处是需要执行的计算量，以及大多数计算都涉及数字处理这一事实。

在20世纪90年代之前，“矢量架构计算机”一直主导着超级计算机的发展。当时，大规模生产的处理器开始变得如此廉价，以至人们可以将它们连接在一起，而不是设计特殊用途的芯片。世界上速度最快的计算机之一是安装在美国国家计算科学中心的克雷XT5美洲豹超级计算机。它有大约1.9万台计算机和22.4万个处理器，它们是基于标准硬件处理器，而不是定制的处理器。

即使是小规模的研究机构现在也可以参与进来，通常是利用商业硬件处理器开发自己的超级计算机。这些计算机被称为贝奥武夫集群。这类计算机是基于现成的处理器，如你在家里的个人电脑中找到的处理器、LINUX操作系统（一种研究人员经常用于科学计算的免费操作系统）以及其他开源软件。

57

贝奥武夫计算机的能力非常强大，而且制造成本低廉：计算机科学教授乔尔·亚当斯和学生蒂姆·布罗姆开发的米罗沃夫集群仅重31磅（小到可以装进手提箱），速度可达26 GFLOPS。这台计算机在2007年初的价格约为2500美元。2009年，世界领先的芯片制造商英特尔发布了一种新的处理芯片，在一个芯片上包含48个独立的处理器。贝奥武夫计算机中使用的这种芯片将为台式机带来真正的超级计算能力。

互联网就像一台计算机

本章首先描述了超级计算机：进行大量计算的机器。这种计算机可以使用数千个处理器。令人惊讶的是，现在有一种更大的计算机，任何使用台式机或笔记本电脑的人都是它的一部分。这就是互联网。

互联网被描述为一个网络，或者更准确地说，是一个网络的网络。让我们看看它是如何运作的。当我从亚马逊等在线书商那里订购一件商品时，我会点击一系列链接，这些链接可以识别我，也可以识别我想买什么书。我每次点击，就会有一条信息发送到亚马逊的一台被称为网络服务器的计算机上。计算机发现我想要的页面，然后把它发回给我，例如包含我最终订单的页面。然后，我点击另一个链接——通常是通知书商我已经完成了我的订单的链接。

作为我的浏览器和亚马逊网络服务器之间交互的一部分，其他的计算机也被使用了。首先，有一些计算机将信息从我的计算机传送到网络服务器——它们不直接传送，而是分成若干个包，每个包可能通过一组完全不同的计算机传送。 58

第二，有一组计算机被称为域名系统（DNS）。DNS是互联网非常重要的一部分。当你输入一个网站的符号名称时，例如http://www.open.ac.uk，DNS会发现这个网站在互联网上的位置。没有这些计算机，互联网将无法运行。

在销售过程中还涉及其他计算机。一个图书零售商可以有许多被称为数据库服务器的计算机。这些包含大量的数据集合；对于在线书商，他们将保存诸如正在出售的图书、它们的价格和库存数量等数据。它们还将包含一些数据，如过去对个人的销售额，以及市场信息，如过去出版图书的销售额。当包含可用细节的网页呈现给客户时，就涉及数据库服务器。

书商的仓库里也会有计算机。这类计算机的一个应用是提供所谓的选书清单：这是在一定时间内所采购书的清单。计算机将这些书分类成一个清单，详细列出每本书及其在仓库中的位置，并将以一种使仓库工作人员所需的行进量最小化的方式来组织这个清单。仓库工作人员还会使用一台计算机来通知数据库服务器图书已经被取走，需要标记为已从书架上取走。

此外，还将出现具有金融功能的计算机，为从图书供应商处购买的图书支付货款，为客户购买的图书发送账单给信用卡公司。因此，这个例子向我们展示的是，互联网的功能就像一系列计算机——或者更准确地说是计算机处理器——执行一些任务，比如买一本书。从概念上讲，这些计算机和超级计算机之间没有什么区别，唯一的区别是在细节上：对于一台超级计算机，处理器之间的通信是通过一些内部电子电路进行的，而对于一组在互联网上共同工作的计算机，通信是通过用于该网络的外部电路进行的。

59

互联网中使用的网络技术可以用来创造一种超级计算机的想法，已经嵌入所谓的网格中，由此产生的计算机被称为网格计算机。

网　格

那么，什么是网格计算呢？计算机网格是使用互联网技术连接在一起的传统计算机的集合，通常通过高速通信电路连接。有两种看待网格的方式：一种是主流超级计算思想的延续，这是研究人员的观点；另一种是优化企业计算机使用的新方法。请记住，计算机处理器在使用时有大量的空闲：它们可以在多达95%的时间里处于空闲状态。商业网格软件供应商指出，购买他们的产品将大大降低企业的硬件成本。

网格可以是正式的也可以是非正式的；前者通常由维护网格的商业软件支持，允许文件共享和处理器共享，后者是执行某些大型任务的计算机松散联盟。非正式网络一个很好的例子是

folding@home。这是一个由斯坦福大学化学系协调的网络。它的目标是进行大量与蛋白质折叠相关的数字运算；这项工作与人类基因组计划有关，它试图找到治疗严重疾病的方法，如帕金森和囊性纤维化。成千上万的计算机以接近4 PFLOPS的速度连接到它。 60

网格计算代表了超级计算机执行大量计算以解决棘手问题的想法，向数字处理领域之外的商业应用程序的微妙转变。在应用领域，它影响了云计算的理念，云计算认为互联网是所有应用的中心工具——而不仅仅是数字运算——并且有可能颠覆目前计算机的商业使用方式。我将在第七章中进一步讨论这个问题。

后　记

本章的内容表明，把计算机想象成放在桌子上的一个盒子，或者是DVD播放器或一组交通灯中的一块硅基电路，这种想法太局限了：对于将互联网看作一个大型计算机，我会在第七章讲述一系列有趣的问题，讨论一种不断发展的商业计算机使用模式，它将大量的处理和数据从单个计算机中带走，并将其委托给由商业企业维护的强大服务器。

这也是我将在最后一章讨论的主题，在这一章中，我将考察乔纳森·兹特林和尼古拉斯·卡尔的作品。兹特林认为，互联网随心所欲的发展使计算机用户进入了一个创新的新时代，但代价是安全等问题。他的著作《互联网的未来》描述了一个可能的场景，商业压力迫使大部分普通计算机用户减少所能做的事情，家庭电脑被降级为类似20世纪60年代的哑终端，那是计

61 算创造力黄金时代结束之时。

卡尔把互联网比喻为计算机，但是集中在工业的观点上。他设想，在未来，计算能力将成为一种公用事业，就像电力成为一种公用事业一样，计算机的作用（至少对房主来说）将降低到
62 只是接入互联网。我将在最后一章再谈到这个主题。

第五章

不安全的计算机

简　介

在2009年伊朗选举抗议活动期间，外国活动人士通过实施所谓的拒绝服务攻击，使作为网络服务器的属于伊朗政府的电脑瘫痪。这次袭击是抗议腐败选举结果的一部分。这些活动分子向服务器发送了成千上万的网页请求，以至于网络服务器的处理器被发送的数据量压垮。这有效地关闭了它们。

1999年，一种名为“梅丽莎”（Melissa）的电脑病毒被发布到互联网上。该病毒所做的是感染微软的电子邮件程序，该程序是视窗操作系统的一部分。该病毒通过电子邮件传播。如果有人收到一封带有病毒附件的电子邮件，然后点击了附件，他们的电脑就会被感染。一旦一台电脑感染了病毒，它就会访问电子邮件程序的联系人列表，并通过电子邮件向该列表中的前50位联系人发送病毒。这是一种特别有害的病毒，不仅因为它传播迅速，还因为它有可能修改受感染计算机上的文字处理文档，

63 从而使它们也传播病毒。

2007年，英国政府报告称，委托给女王陛下的税务和海关部门的超过2500万人的儿童福利数据丢失。详细资料储存在两张光盘上，并邮寄到另一个部门。

这只是计算机不安全的三个例子。前两项涉及技术故障，而最后一项则是管理故障。本章的目的是考察计算机面临的威胁，以及如何应对这些威胁。它还将探讨一些错综复杂的隐私问题。

病毒和恶意软件

病毒是非法引入计算机或计算机网络的计算机程序；一旦引入，它就会进行一些恶意行为。典型的恶意行为包括：删除文件；就梅丽莎病毒而言，通过电子邮件将文件发送给其他互联网用户；监控用户的按键操作，以发现密码和银行信息等重要信息；扰乱文件，使其无法阅读，然后要求支付赎金，当受害者支付赎金时，会向受害者发送一封电子邮件，说明如何还原文件。病毒是被称为恶意软件的软件集合的一个子集，而且是一个很大的子集。

病毒进入计算机系统的主要方式有三种。第一种方式是作为附件，例如，用户可能会收到一封电子邮件，其中包含一条关于名人惊人行为的消息，并指示用户点击附件以查看该行为的照片证据；一旦收件人点击附件，病毒就会在计算机上驻留。64

计算机感染病毒的第二种方式是使用不安全的软件。互联网上有很多免费软件；大部分都是有用的（我使用一个优秀的程序来备份我的文件），但是有些软件会包含病毒。一旦安装了

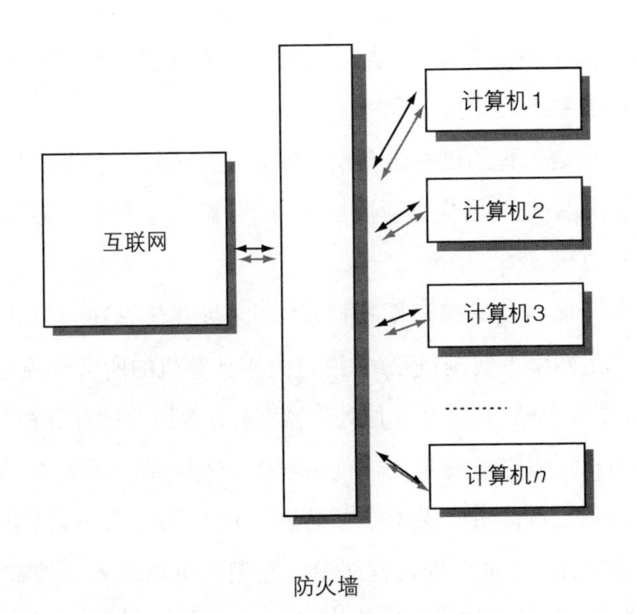

图6　防火墙结构

包含病毒的软件，病毒也会被安装。病毒携带者通常被称为“特洛伊木马”。

第三种方式是不使用或不正确使用被称为防火墙的程序。这是一个屏障，可以防止不必要的连接进入计算机。它会持续监控这些连接，并拒绝任何被认为有害的连接。

除了病毒和特洛伊木马之外，还有许多其他类型的恶意软件。逻辑炸弹是一种程序，当特定情况发生时，它会在计算机中执行，比如说当一个文件被首先访问的时候。定时炸弹与逻辑炸弹类似，只是恶意软件的执行发生在未来的某个时间。这类炸弹主要被离开公司的不满员工所使用。 65

陷阱门是计算机系统中的一个弱点，允许未经授权进入系统。陷阱门通常与写得不好的系统联系在一起，这些系统与计算机的操作系统交互，但也可以由那些为计算机受到影响的组织工作的系统程序员故意构造。

蠕虫是一种通过网络多次复制自身的程序。与互联网相关的首批入侵之一是莫里斯蠕虫病毒，它影响了新兴互联网中的大量计算机。

兔子病毒与蠕虫病毒相似，不同之处在于它在一台计算机上而不是网络上复制自己，并通过占用计算机的所有资源（主存储器、文件存储器，甚至硬件处理器）来有效地关闭计算机。

病毒可以隐藏在多个位置，例如，它们可以与程序集成：它们可以包围程序并拦截对要执行程序的任何请求，可以附加到程序的末尾，也可以嵌入程序中。它们还可以嵌入文档或图像中，然后在打开文档或图像时执行。

用于检测、隔离和删除病毒的一项重要技术是病毒扫描仪。诺顿、卡巴斯基和AVG等公司提供了大量商用和免费的病毒扫描程序。病毒只是一个程序——尽管是一个可能造成重大破坏性影响的程序——和所有程序一样，都有一个签名。这是病毒特有的0和1模式。病毒扫描器在执行时会处理计算机的硬盘和存储器，寻找病毒特征。为了做到这一点，它依赖于已知病毒特征的数据库。

当你购买病毒扫描器时，你也会购买一个病毒特征数据库和一个程序，当出售扫描器的公司发现并检查新病毒时，该程序会定期更新数据库。

查尔斯·P.弗莱格和莎丽·L.弗莱格在他们关于计算机安全的优秀著作中描述了保护自己免受病毒侵害所需采取的步骤：使用病毒扫描器；仅从信誉良好的商业软件供应商处购买

软件；在一台孤立的计算机上测试你怀疑的所有新软件；如果你收到一封带有附件的电子邮件，那么只有在你知道它是安全的情况下才能打开它（一项检查是在谷歌等搜索引擎中键入电子邮件的一些关键词：你可能会发现有人将该电子邮件标记为危险邮件）；而且要经常备份你的文件，比如把它们转移到一个便宜的硬盘上，而这个硬盘通常与你的电脑分开存放。

之前已经简要提到的另一种技术对病毒和其他攻击非常有效，那就是防火墙。有各种各样的防火墙架构，一种架构如图6所示。防火墙所做的是拦截来自受保护网络（通常来自互联网）之外的任何流量，并拒绝任何可能对网络造成危险的流量。还有一种被称为个人防火墙的防火墙，用于保护个人电脑；然而，它的功能类似于工业防火墙：只是它有一个较小的功能集。这种网络防火墙通常是为家庭使用而销售的，并作为一个包含病毒扫描器和其他安全软件的集成包销售。

防火墙执行的典型操作包括：检查任何附件是否包含病毒，拒绝访问计算机或网络，以及拒绝向外部计算机提供设施，例如复制文件。

计算机犯罪

有大量与计算机相关的犯罪，包括：通过操纵计算机记录来进行欺诈；在一些国家，垃圾邮件是非法的，即发送未经请求的电子邮件，通常销售某些产品或服务；未经许可访问计算机系统；未经许可访问计算机系统以读取、修改和删除数据；窃取数据或软件；通过计算机进行的工业间谍活动；通过访问某人的个人信息窃取其身份；访问财务数据并窃取与该数据相关的资金；传播病毒；传播儿童色情制品。

一种典型的非技术性犯罪是，员工可以访问金融系统，并签发大量应付给自己和/或他人的支票。在犯罪者被抓获之前，此类犯罪通常会涉及超过100万美元的资金。

2008年，一个包含翻译广告的网站发布了一个职位空缺信息。这项工作包括修改从另一种语言翻译成英语的文本。申请者被要求提供简历中常见的数据，以及他们的银行账户详细信息。发布该广告的"公司"与乌克兰的一个犯罪集团有关。任何提出申请的人都会被抽走账户中的资金，然后用于洗钱骗局。

2009年，一项针对美国计算机的调查发现，美国有数百万台计算机被假冒安全软件"感染"。这是一个非常复杂的骗局的结果。在这个骗局中，当计算机用户访问某个特定网站时，他们在弹出窗口中看到一个警报，告诉他们自己感染了病毒，可以轻松下载免费软件，并与其他许多软件一起用于检测和删除病毒。当然，该软件是一种能够发现计算机用户财务细节的病毒。

2006年，三名俄罗斯黑客因对英国博彩公司实施了一次拒绝服务攻击而被判入狱八年。在这种类型的攻击中，计算机中充斥着消息——有时是网页请求，有时是电子邮件消息——结果是，受到攻击的计算机被禁用，无法响应正常用户的请求。黑客攻击的目标是一家大型在线博彩公司的网络服务器，该公司因拒绝支付巨额赎金，他们的计算机在"育种者杯"比赛期间被封锁，致使该公司每天的攻击损失超过16万英镑。

关于计算机犯罪的一个重要观点是，许多犯罪可以由缺乏技术知识的计算机用户实施。上面的例子需要不同程度的技术技能，而抽走资金的例子则需要最少的技能。典型的非技术性犯罪包括：在大型垃圾箱中翻找含有敏感信息的计算机打印件（这一过程被称为垃圾搜寻），用手机拍摄白板上写的有关计算机系统的重要信息，窃取密码并使用该密码伪装成计算机系统的用户。

所有这些非技术性的例子最多需要你通过使用家庭电脑进行文字处理或电子表格工作所获得的知识。因此，考虑到计算机犯罪可以用技术或非技术的方式进行，对计算机犯罪有各种技术和非技术上的防御也就不足为奇了。 69

计算机犯罪的技术性防御

针对大量与计算机相关的技术犯罪的主要防御措施之一是密码学。这是一种用于隐藏和加密数据的技术，这些数据可能存储在计算机中，也可能在通过网络连接的两台计算机之间传输。密码学在罗马时代就已经存在了，当时尤利乌斯·恺撒用它向他的军事指挥官发送信息。计算中使用的密码相当复杂；不过，对恺撒密码的描述会让你大致了解它们的工作原理。

恺撒密码涉及通过将信息中的每个字母替换为字母表中该字母后面的第n个字母来更改文本。例如，如果n是2，那么字母“a”将转换为字母“c”；字母表末尾的字母将转换为开头的字母，例如字母“y”将转换为“a”。

Caesar was here

在这里将被改成

Ecguct ycu jgtg

在恺撒密码中，数字n充当密码。在密码学界，它被称为密钥，即用来控制加扰文本（纯文本）的精确转换，以产生无法读取的文本（密文）。上述信息的接收者所需要做的就是知道密钥（每个字母向后转换位置的数字）。然后，他们就能从密文中提取信息。

恺撒密码非常简单，也很容易破解。破解现代密码是一项复杂的工作，需要掌握统计学知识才能理解。它需要关于词频、字频的数据，理想情况下，还需要一些关于电子邮件上下文和消息中可能包含的一些单词的知识。正是对信息中可能出现的短语的了解，使得第二次世界大战的密码学家能够破译德国信息。例如，许多信息都会包含“Heil Hitler”。

自恺撒密码问世以来，我们已经走过了漫长的道路；事实上，自第二次世界大战以来也已经有很长时间了。基于计算机的密码仍然是基于替换，例如恺撒密码中的类型。然而，有两个因素使得密码变得非常强大。首先，换位既适用于纯文本，也适用于替换；一种换位是指文本中的一个字符移动到文本中的另一个位置。第二，大量的替换和换位通常通过计算机硬件或高效的计算机程序进行。

我所描述的密码类型被称为对称密码；它是对称的，因为更改纯文本（加密）和将密文转换回纯文本（解密）使用的密钥相同。对称加密方法非常有效，可用于对从一台计算机发送到另一台计算机的大型文件或长消息进行加密。

不幸的是，对称技术面临一个主要问题：如果有许多人参与数据传输或读取文件，每个人都必须知道相同的密钥。这让它成了一场安全噩梦。更糟糕的是，密钥越复杂，消息被读取的可能性就越小；因此，像“Darrel”这样的密钥是无用的，而像“55Tr3e343E4((!!dfg21AVD90kk-9jk}”这样的密钥则提供了高度的安全性，唯一的问题是它可能很难被记住，需要存储在某个地方，例如钱包或桌子的抽屉中。 71

虽然已经有很多人试图确保密钥分发过程的安全性，例如，对生物识别智能卡上的密钥进行编码，并为密钥或密钥集的分发设计安全协议，但对称加密和解密的密钥安全性仍然是一个问题。

这个问题的一个公开解决方案是由三名美国计算机科学研究人员开发的，他们是惠特菲尔德·迪菲、瑞夫·墨克和马丁·赫尔曼；而使他们的想法得以实际应用的则是另外三名研究人员，他们是罗纳德·李维斯特、阿迪·萨莫尔和伦纳德·阿德尔曼。我使用“公开”这个词，是因为英国政府解密的文件表明，三名英国研究人员——克利福德·柯克斯、马尔科姆·威廉森和詹姆斯·埃利斯——也在英国政府位于切尔滕纳姆的绝密通信总部开展了这方面的工作。他们的工作是在20世纪70年代初进行的。

这些研究人员开发的解决方案，被称为公钥加密或非对称加密，其核心在于计算机无法解决需要大量计算资源的难题。术语“不对称”最能描述这种技术，因为它需要两个不同的密钥，一个公钥和一个私钥。

假设两个计算机用户A和B希望使用非对称加密技术进行通信，并且每个用户都有一个公钥和一个私钥。公钥由两个用户各自发布。如果A想向B发送加密消息，那么她将使用B的公钥对消息进行加密。当B收到消息时，他将使用他的私钥解密信 72

息。非对称密码系统有许多特性；一个非常重要的问题是，用于发送消息的公钥不能被某个截获消息并试图解码它的人使用。

公钥密码术一举解决了与对称密码术相关的一个主要问题：存在大量密钥，其中一些密钥可能以不安全的方式存储。然而，非对称加密的一个主要问题是，它的效率非常低（大约比对称加密慢一万倍）：虽然它可以用于电子邮件文本等短消息，但对于发送千兆字节的数据来说，它太缺乏效率了。然而，正如你稍后将看到的，当它与对称加密相结合时，非对称加密提供了非常强的安全性。

一个强大的用途是提供一种数字身份形式作为数字签名，其中密钥生成过程的数学特性用于提供认证，证明声称发送了消息的人实际上就是该人。

第二个用途是发送安全消息，例如在银行和客户之间。在这里，用户和网络服务器的身份通过数字身份检查进行介导。这里值得一提的是，你不应该仅仅依靠技术，而应该使用常识。例如，一种常见的欺诈行为是向用户发送一封自称来自银行的电子邮件，要求他们登录一个伪装成银行网站的流氓网站，谋取用于抽走资金的账户数据。互联网上到处都是颠覆科技的恶棍。

有许多技术被用来提供这种安全性；几乎所有的密码都是基于对称和非对称密码的组合。一种非常流行的安全方案被称为安全套接字层，通常简称为SSL。它是基于一次性密码本的概念。它提供了一个近乎理想的加密方案。它要求发送方和接收方都有一个名为密码本的文档，其中包含数千个随机字符。发送方取前*n*个字符，比如50，然后使用它们作为密钥对消息进行加密。接收者在收到信息时，会从他们的密码本中取这50个

73

字符并解密信息。一旦这些字符被用于密钥，它们就将被丢弃，并使用接下来的*n*个字符发送下一条消息。

一次性密码本的一个主要优点是，一旦密码被使用就会被丢弃；另一个优点是，任何文档都可以用来生成密钥，例如电话簿或小说也同样适用。一次性密码本的主要缺点是，它们需要发送方和接收方之间同步，以及密钥的打印、分发和存储的高度安全性。

SSL使用公钥加密技术在消息的发送方和接收方之间传递随机生成的密钥。该密钥对于发生的数据交换仅可使用一次，因此是一次性密码本的电子模拟。当交换的每一方都收到密钥时，他们使用对称加密技术对数据进行加密和解密，生成的密钥执行这些过程。如果双方希望进一步交换数据，则会生成另一个密钥，并像以前一样进行传输。

针对计算机犯罪的另一项技术防御措施是密码。密码背后的理论很简单：计算机用户记住一些字符集，并使用它们访问网站或文件——密码实际上是用户身份的代理。有时密码被用作加密密钥。密码提供了主要优势；然而，由于人为失误，密码方案的安全性可能会降低。

一般来说，计算机用户选择的密码很差。最好的密码和加密密钥是长密码且包含字母、数字和特殊字符，例如叹号。举例来说，密码"s22Akk;;!!9iJ66s0-iKL69"是一个很好的密码，而"John"是一个很差的密码。20世纪90年代，埃格网上银行发现其客户选择了糟糕的密码。例如，50%的人选择了家庭成员的名字。查尔斯·P.弗莱格和莎丽·L.弗莱格在他们出色的《计算机安全》一书中就密码选择提供了明智的建议：不要只使用

74

字母字符，使用长密码，避免使用姓名或单词，选择一个不太可能记住的密码，例如“Ilike88potatoesnot93carrots!!”，定期更改密码，不要写下密码，也不要告诉别人你的密码是什么。

除了密码，还有许多高科技的方法可以在电脑上识别你自己。生物识别是一个备受关注的领域。在这里，使用的技术可以从独特的身体特征识别计算机用户，通常与密码结合使用。可以使用的典型特征包括指纹、虹膜图案、用户的脸、他们的声音以及一只手的几何形状。许多技术都处于研究阶段，有些技术的错误率高得令人无法接受，而且容易受到攻击。例如，语音技术的一个活跃领域是从一组录音中模拟一个人的声音。

非技术性安全

在我从查尔斯·P.弗莱格和莎丽·L.弗莱格的著作《计算机安全》中引用的建议中，有一条禁令是禁止告诉任何人你的密码。这是预防非技术性安全漏洞和问题的一个很好例子，因为计算机用户面临的不仅仅是与病毒和其他非法技术有关的问题。

75

前安全黑客凯文·米尼克在其著作《欺骗的艺术》中描述了一些对计算机系统的攻击，这些攻击完全不需要技术知识。许多黑客已经使用了各种被归类为社会工程的技术来渗透计算机系统——不是通过使用蠕虫病毒或操作系统陷阱门，而是通过利用人性的弱点。他在书中指出，社会工程陷阱

> ……利用影响力和说服力来欺骗人们，让他们相信社会工程师不是他，或者通过操纵。因此，社会工程师能够利用人们在使用或不使用技术的情况下获取信息。

这是米尼克描述的一个典型的社会工程攻击。潜在入侵者通过伪装成另一家商店的店员，与音像租赁商店的店员建立关系，该商店是一系列音像商店的一部分。这种关系持续了几个月，以至于无辜的店员认为与他们正在交谈的人实际上在姐妹店工作。然后，有一天，入侵者打电话给店员，声称他们的电脑坏了，并询问音像租赁客户的详细信息（姓名、地址、信用卡）。一旦获得了这些信息，他们就可能从客户的银行账户盗取资金。

除了使用非技术性手段犯罪之外，数据也因人性的脆弱而丢失。过去的十年中，公司和组织出现了大量数据泄露。例如，2008年，英国汇丰银行承认丢失了数万名客户的保险记录数据。虽然磁盘受密码保护，但没有加密，因此任何具有良好计算技能的人都可以读取。

76

2007年，在英国计算机史上最大的一次数据泄露事件中，政府的税收和海关部门丢失了2500万份记录，其中包括那些领取儿童福利的英国公民的财务和其他数据。一名员工将详细信息写入计算机磁盘，并将未登记的磁盘寄送给另一个政府部门，然后在邮寄过程中丢失。

公众的印象是，对安全和隐私的主要威胁来自技术性攻击。然而，来自更平常原因的威胁也同样高。数据盗窃、软件和硬件损坏以及未经授权访问计算机系统，可能以各种非技术性方式发生：有人在垃圾箱中查找计算机打印输出；由使用手机摄像头的窗户清洁器拍摄包含敏感信息的显示屏；办公室清洁工从办公桌上偷文件；公司访客在白板上记下密码；一名心怀不满的员工用锤子敲打一家公司的主服务器和备用服务器；或者，有人把未加密的记忆棒扔到街上。

保护计算机免受技术性攻击相对容易：它涉及购买安全软件，并确保该软件使用的文件（例如病毒特征文件）是最新的。当然，使用技术性手段访问计算机的入侵者和负责安全的工作人员之间总是存在斗争。有时攻击会成功：例如，20世纪90年代的拒绝服务攻击是一个严重的问题，但很快，技术性手段就会被开发出来以应对它们。

对于公司和其他组织来说，防范非技术性攻击要困难得多：它需要一整套程序来防范所有可能的安全风险。此类程序的例子包括：接待访客的程序，例如执行不得访客独自在建筑物周围走动的规定；废弃的计算机打印输出的处理程序；确保笔记本电脑在旅行时的安全；硬盘上可能存储敏感数据的旧电脑的处理程序；禁止在社交网站上发布可能有助于社会工程攻击者访问计算机网络的个人信息；以及确保桌面清洁政策。

77

非技术性安全要困难得多，因为它比技术性安全要普遍得多：它需要从接待区工作人员到计算机中心负责人的所有人的合作，而且，在重视安全性的组织中，它被嵌入一个厚厚的文件中，称为安全手册或安全程序手册。

对于在家工作的个人来说，想要防范非技术性攻击要容易得多，所需要做的就是：加密任何敏感文件，例如包含财务数据的电子表格；为任何敏感数据文件使用足够安全的密码；切勿在互联网上或接听电话时提供密码和银行账号等数据；如果你确实需要随身携带数据，那就买一个加密的记忆棒——它们过去有点贵，但自首次上市以来价格下降了许多。

78

第六章

破坏性计算机

鲍德斯英国公司

今天早上，当我开始写本书第一章的初稿时，就听说我最喜欢的书店之一鲍德斯英国公司陷入了财务困境——四天后，它们进入了破产管理状态。英国广播公司的新闻谈及其中的一个原因是，该公司发现很难与亚马逊等在线零售商竞争。经营连锁书店的成本要比仓库、网站和呼叫中心贵得多，其中一些书店位于黄金购物地点。作为比较，我看了一下我最近买的一本书的价格。在市中心的一家商店里，我会以34英镑的价格买到它，而亚马逊网站上目前宣传的价格是20.13英镑。

虽然鲍德斯公司的员工能够使用计算机来帮助他们回答客户的问题和订购缺货商品，但最终它们威胁到了他们的工作。我确实认为书店有着某种未来，因为它们提供了浏览的设施，但直到我看到许多书商通过一个被称为谷歌预览的系统，允许访问其网站的访问者浏览一本书的目录和书中的许多页面。对于

79 图书交易来说，计算机具有破坏性。

我在米尔顿凯恩斯最喜欢的唱片店是维珍大卖场，在被管理层收购后，它改名为“扎维”（Zavvi）。此后不久，扎维零售连锁店倒闭。其中一个原因是，从iTunes等网站可以下载更便宜的音乐。扎维的工作人员再次发现，计算机在他们的工作中很有用，但最终它扼杀了他们的生意。对于音乐行业来说，计算机具有破坏性。

我可以从计算机上阅读全国性的报纸；我偶尔会这样做，但大多数时候我都会访问英国广播公司新闻网站，该网站因其设计、执行和内容获得了多个奖项。互联网上新闻的日益普及对报纸的销售产生了巨大的影响。西方世界的报纸一直在应对广告收入下滑、发行量下降以及读者在线免费新闻运动。根据美国发行量审计局的数据，与2008年同期相比，2009年4月至9月，379份美国日报的平均日发行量下降了10.62%。计算机帮助了那些为报纸复制文件的记者：文字处理软件是一款非常棒的软件。然而，计算机导致许多报社裁员。对于报业来说，计算机具有破坏性。

与报纸相关的另一个破坏性是评论（例如书评）空间的减少，以及评论人员（包括长期的和特约的）往往是第一个感受到裁员寒风的人。书评人和影评人过去是非常有影响力的人物：他们的观点可能会扼杀一本书或一部电影，或者在热门列表中提升它。现在有很多网站都在评论电影和书籍。它们中的一些只专注于评论——比如“烂番茄”网站——或者把提供评论作为其业务的一部分：这里的一个很好例子是亚马逊网站，尽管它

80 致力于销售各种商品，但也涵盖商品目录中的商品评论——由

客户撰写的评论。

在本书的这一部分中，我将从我们如何与他人互动、就业以及技术如何改善我们的生活等方面，来探讨计算机对我们的影响；我还将探讨一些可能对我们产生负面影响的方式。

破坏性技术

克莱顿·克里斯滕森是关于技术的破坏性影响的重要作家之一。他的《创新者的困境》和《创新者的解决方案》两本书探讨了硬盘存储单元等技术设备如何对行业产生重大影响。

破坏性创新可分为两类：低端破坏性创新和新市场破坏性创新。后者是指技术进步创造了新的商业机会，改变了消费者的行为，并往往导致工业子部门的消失。低端破坏性创新会影响当前的技术对象和服务，降低它们的价格，从而降低它们的可用性。

计算机已经导致了这两种类型的破坏性。例如，当手机被设计和开发时，生产它们的公司在事后添加了信息功能，没有想到这会是一个使用如此之多的功能，以至于现在无法想象在没有短信功能的情况下销售手机。这是一个新市场破坏的例子，它摧毁了大部分的寻呼机行业。

低端破坏性的一个例子是开源软件，我将在本章后面讨论。这使得计算机用户能够使用操作系统、文字处理器、电子表格系统和照片实用程序，而这些都不需要任何成本——这是真正的降价。

81

开源开发

当你使用计算机访问互联网时，你可能有数亿用户的受众，并且有可能将这些用户联系在一起。维基百科是一个大规模合作的例子：所有条目都是由志愿者发起和编辑的。另一个同样令人印象深刻的例子是开源软件的兴起。这种软件在许多方面都不同于商业软件；首先，它是免费的；其次，开源软件的程序代码可以供任何人阅读，这就是“开源”一词的由来，“源”指的是程序代码；最后，任何人都可以获取代码和修改代码，甚至将其作为产品销售。

存在大量开源程序代码，最常用的两种是Apache和Linux。前者是用于从网络服务器分发页面的软件。目前，估计大约65%的网络服务器使用Apache。Linux是一个更令人印象深刻的故事。它是一种操作系统，是流行的视窗操作系统的竞争对手。它起源于一个名为MINIX的简单操作系统，该系统是由计算机科学学者安德鲁·塔南鲍姆开发的，目的是教他的学生如何设计大型系统。

Linux是由赫尔辛基大学的学生林纳斯·托瓦兹开发的。他决定改进MINIX并开始开发Linux，最初的灵感来自MINIX的一些设计思想。Linux最初的实现非常好，在互联网蓬勃发展的同时，它引起了许多软件开发人员的注意。其结果是，该系统逐渐演变为由付出时间而不求回报的程序员来维护。

82

Linux在渗透性及其所提供的功能方面都令人印象深刻。最近的一项调查显示，80%最可靠的网络托管公司使用Linux，它是许多超级计算机的首选操作系统。还有一个桌面版的Linux，它包含了你所期望的一系列常用软件：文字处理器、图形程序、图像编辑器、电子邮件程序和音频播放器。

在桌面方面，Linux距离威胁非常流行的视窗操作系统还有一段路要走。例如，为桌面版本安装新软件有时很麻烦。但这仍然是计算机用户在现有机构之外创造大型复杂人工制品的趋势中最令人印象深刻的例子。

广　告

计算机对电视公司和报社的收入产生了重大影响。这主要是因为使用了在线广告，但也因为包含广告的电视节目可以被录制下来，当回看时，广告可以被快速转发。在这些行业发生的破坏性关键是目标市场选择。这个概念并不是什么新鲜事：它意味着你把广告放在你认为它们会获得最大读者群和回报的地方。例如，如果你有一则关于希腊考察团的广告，其中包括参观古遗址和聆听著名希腊古典主义者的演讲，那么你可能会把广告定向投放在《独立报》这样的报纸。

计算机和互联网改变了目标市场选择的面貌，并使其更加有效。具有破坏性的主要技术是“关键词广告”，这是一项谷歌的技术，它为该公司赚取了数十亿美元。“关键词广告”基于点击付费模式。如果广告主希望在谷歌上为自己的产品或服务做广告，他会指定一个金额，当某个人进行谷歌网络搜索时，一个在线广告出现在检索到的页面上，他就要支付该金额。

广告主指定哪些关键词会触发广告，因此，如果用户在谷歌搜索框中键入“钓鱼”一词，他们会发现——正如我键入该词时所看到的那样——显示了三个在线钓鱼网站的广告、一个钓鱼假期的广告和一个亚马逊钓鱼项目的广告；还有一个网站链接，该网站介绍了一种将鱼饵扔进河流或湖泊的革命性方法，而另一个链接的网站则收集了与钓鱼相关的链接。

与传统媒体相比，“关键词广告”代表了一种更具细粒度的广告方式。它效果显著。2009年第三季度，美国报纸的广告收入与2008年同期相比下降了28%；2009年前九个月的收入也下降了29%。显然，这种下降的部分原因是美国企业近年来经历的信贷紧缩问题；然而，这也是21世纪初以来出现的一种重要趋势。

电视行业的收入也有所下降。例如，根据《泰晤士报》的分析，谷歌在英国的收入为3.27亿英镑，而2007年7月至9月间，英国商业频道独立电视一台的全部产值为3.17亿英镑。

信息技术外包

2011年，我的英国电信互联网连接突然中断；对我来说，这相当于断电或停水（当我看尼古拉斯·卡尔的作品时，本书最后一章讨论了这一点）。我拨通了英国电信的求助电话，一个印度口音的人接听了电话。在接下来的二十分钟里，他通过互联网控制了我的电脑。他打开文件夹，改变了一些设置，重启了一些程序，瞧，我的连接重启了。我不经常激动，但这次我很激动。我看着鼠标指针在一只看不见的手的引导下在屏幕上移动——当我开车经过希思罗机场，看到一架波音747型客机起飞时，我会有一种感觉：我知道我所看到的是可能的，但并不完全相信它。明确地说，过错不在英国电信：我习惯于打开和关闭互联网调制解调器，在关闭期间，它错过了一个重要的更新。

84

这是外包的一个例子。在许多传统行业，外包已经成为常态。例如，在英国销售的大部分服装都是在中国、印度和多米尼加共和国等国生产的；电子设备通常是在劳动力成本较低的经济体中生产的，比如我的iPod是中国制造的。然而，外包现在在系统开发中也很常见。

英国领先的律师事务所之一品诚梅森律师事务所，列出了使用外部软件开发人员的理由：效率和规模经济带来的成本降低；获得高水平的信息技术技能（例如，印度的软件开发人员是世界上最精确、最老练的程序员之一，他们使用先进的工具进行系统开发；他们还拥有最高比例的经认证可以生产最高可靠性软件的公司）；从公司的基础设施中移除非核心业务；最大限度地减少信息技术基础设施的巨额资本支出，并且对未来的支出有一定程度的把握。

互联网提供了一个基础设施，使客户能够通过视频链接与其他国家的系统分析师交流，通过电子邮件发送文档，并通过实时链接测试系统。其结果是，基于计算机的开发越来越多地被转移到海外。例如，2006年，研究公司“计算机经济学”的报告称，在所有接受调查的美国公司中，61%的公司将部分或全部软件开发外包。

当我坐着敲打这一章时——我们的金融体系经历了一场最大动荡的三年后——信息技术外包的未来并不明朗。一方面，公司正在削减信息技术投资；另一方面，离岸外包公司在投资方面提供了显著的节省。这两个因素，再加上信息技术技能水平的提高，将决定未来几十年外包的增长。

我上面讨论的外包类型是粗粒度的外包，即你雇一家公司来完成一些信息技术功能。还有一种细粒度的版本。现在有许多网站提供软件员工的详细信息（这些程序员通常来自东欧）。因此，如果你有一个项目需要在有限时间内进行少量的软件开发，那么视频链接和电子邮件可以让你接触到与西欧价格相比具有竞争力的开发人员。

公民新闻

“公民新闻”一词用来描述普通公民如何通过廉价计算机和互联网连接，参与报道对事件的反应，发表新闻文章和观点文章。有许多技术可用于此：主要的技术有博客（在线日记）、播客、视频片段、数码照片和网站。

这种现象有许多表现形式。有些新闻博客聚合新闻并包含评论，这些评论不仅来自博客作者，还来自其他互联网用户，他们可以在博客文本的末尾插入评论。这些博客的内容可以是一般性的，也可以是特定主题的，比如技术。

还有一些新闻网站包含与传统报纸网站相同类型的材料。其中一些网站持中立观点；然而，他们经常从特定的立场报道和评论新闻，例如从绿色运动的角度。

公民新闻的一些最有趣的表现形式是参与式新闻网站，在这些网站上，计算机用户给文章回帖，其他用户用描述符标记文章，提供简单的索引，在掘客等网站上，用户投票决定哪些文章有趣。回帖数量最多的文章会被提升到网站的显著位置。

计算机和硅基电路价格下跌的结果之一是，数字设备的价格和可用性随之下降。其中一个最显著的领域是数字录音。在我的左边，当我写这一章时，我有一台由马兰兹公司制造的数字录音机。它的价格约为400英镑，音质可与电台采访人员使用的录音机媲美，且只有其成本的一小部分，比十年前的磁带录音机更方便。任何人都可以花800英镑左右买到两个电容话筒、一个便宜的混音器和一个这样的录音机，还可以把自己变成一个互联网电台。很多人都可以这样做。

计算机和互联网为自由表达提供了一种媒介，直到21世纪初，这种媒介才为记者所用。之前任何人能想到的最多就是写一封信给编辑，这封信可能会被选中，也可能不会被选中。现在，只要有一个重大新闻，互联网就会被消息淹没。

这种影响的一个例子发生在2009年11月。黑客成功地访问了东安格利亚大学气候研究所使用的计算机。这是世界上最重要的全球变暖研究机构之一。几天之内，互联网上就流传着大量电子邮件、音频播客、视频播客、博客和新闻故事。在独立审查调查这些说法时，该部门主管选择了辞职。

此事件发生五天之后，我在谷歌上搜索“气候门”这个词——气候怀疑论者用这个词来形容这起事件，点击量超过1300万次。我还查看了“优兔”视频网站，当我输入相同的关键词时，点击量超过1.6万次。

很明显，计算机以一种直接方式（传播新闻的能力）和一种间接方式（缺乏安全性）已经并将对新闻业产生重大影响。

数码摄影

有很长一段时间，我都用胶片相机。我会在当地的药店或摄影店买一卷胶片——通常有足够的空间进行36次曝光——然后把它装进相机，拍下照片，最后把曝光的底片交给药剂师冲洗。之后，药剂师会把胶片送到一个正在开发的实验室，最后，我会收到我的“快照”。业余爱好者有时会冲洗自己的照片；为此，他们需要将胶片浸泡在各种化学物质中，然后将底片放入被称为放大机的设备中。这会在一张浸有化学物质的纸上产生一个图像。之后，再在含有更多化学品的若干托盘中对纸张进行冲洗。这一切几乎都是在暗室中发生的，只能用红灯。

我现在用数码相机拍照，在我所使用的环境下，可以拍几百张照片。我需要做的就是把它们加载到计算机中，使用一系列可用的图像查看器和操作程序中的一个。

一个业余爱好者想要修改基于胶片的图像，就必须进行一些复杂的、容易出错的操作；例如，他们必须在放大机光束和照片之间转变纸板形状，以限制光线，从而改变图像的光线值。Photoshop等现代照片处理程序提供了各种工具，使摄影师能够以多种方式处理照片，例如改变颜色，改变照片整体或部分的曝光，着色，创造类似于后印象派画家的效果，然后在照片上选择区域并放大。

2002年，数码相机的销量超过了胶片相机。当我去一个旅游景点，现在很少看到正在使用的胶片相机。从那时起，数码相机的发展趋势是设备越来越好，比如分辨率越来越高。这种变化是由于硬件改进、更快的处理器、图像传感器像素密度的增加以及电路尺寸的减小。摄影软件提供的工具也有所增加。例如，有一种被称为高动态范围成像的技术，可用于生成超逼真的照片，这些照片是使用不同曝光值拍摄的数字照片的多个版本组合而成的。

数码相机也有被嵌入其他设备的趋势，主要是手机。移动电话的增长令人震惊。2002年，全世界约有10亿移动电话用户和10亿固定电话用户。2008年，约有12.7亿固定电话用户和40亿移动电话用户。西方世界的绝大多数手机都配有摄像头。这样做的一个后果是，它创造了一场公民记者运动。例如，贾尼斯·克鲁斯在哈德孙河拍摄了美国航空公司1549航班的第一张照片，因为引擎问题，客机在水上迫降。客机坠入水中之后，他用iPhone拍摄了这张照片——一张引人注目的照片显示，乘客们站在客机的一个机翼上，其他人则蜷缩在漂浮的紧急降落伞上。2009年6月，伊朗发生了大规模抗议活动，称总统选举被操纵；许多抗议活动被当局暴力镇压。几分钟后，互联网上就可以看到反映此事件的照片，因为抗议者将用手机拍摄的照片发送给了伊朗境外的朋友。

数码摄影的兴起还有许多其他影响。一个明显的问题是化学开发实验室的消亡。计算机的日益强大及其日益小型化意味着，这些实验室没有简单的升级途径，无法将自己转变为数字实验室，因为当地的摄影商店或连锁药店现在可以投资于复杂的数字摄影设备——现在只剩下几家非常专业的公司，它们迎合了网络艺术行业的需求，例如拍摄非常细腻的黑白照片的摄影师。

还有一些不太明显的影响。一些文化评论员提出了对照片的信任问题。例如，政治对手在一次反越战集会上散发了一张总统候选人、参议员约翰·克里和简·方达的数码照片——这张照片是经过篡改的。有许多机构在照片中插入了黑人面孔，以给人一种更具种族多样性的印象。

所有这些都提出了我们可以信任哪些图像的问题。显然，这种处理可以用胶片技术来完成，但要做到这一点就不那么容易了。使用计算机程序对数字图像进行扫描则要容易得多：我花了四分钟就将我妻子在我们村子里拍摄的照片转换过来，并添加了在阿维尼翁拍摄的背景，以给人留下她陪我旅行的印象。

89

90

除了使用手机，数码摄影还有另一种公民新闻效应。阿布格莱布事件起因于向朋友和熟人发送美国士兵虐待伊拉克囚犯的数码照片。

数码相机也提供了更多的创意机会。约翰·西蒙、肖恩·布里克西和帕斯卡·多姆比斯等艺术家在工作中经常使用计算机。得克萨斯大学的德克·霍尔斯特德教授对该杂志的读者进行了一项调查，这位数字记者向读者询问了他们对数码相机技术的态度。他们都喜欢数字技术；这项调查的一个令人惊讶的结果是，它提供的创造力比速度和便利等因素更具优势。

科学与计算机

科学研究中发生的一件令人震惊的事情是，它正被大量的数据所淹没。计算机现在经常被用来存储和分析来自地球轨道卫星、陆地气象站和大规模核试验的数十亿条数据。

例如，在未来十年内将投入使用的大型巡天望远镜，将使用非常复杂的计算机硬件，与数字摄影中使用的硬件类似，在运行的第一年，它将产生大约1.3拍字节的数据，比任何其他望远镜提供的数据量都大。

这一系列数据为科学家提供了分析和预测的重要机会，例如预测强降雨对主要人口中心附近洪水的影响。然而，它也带来了一些重大问题。世界各地有大量不同的研究小组提供数据。如果一个研究小组，比如大气物理学，想要来自另一个研究小组的数据和代码，正常的过程是发送电子邮件，如果另一个研究小组想要相同的数据，则会发生相同的过程。因此，科学家面临的第一个问题是分布式数据。

91

第二个问题是没有可接受的存储标准。有两个组件与数据存储相关。首先是数据本身：由空格或逗号等符号分隔的大量浮点数。其次是元数据。这是描述原始数据的数据。例如，如果数据来自全球范围内的一系列温度测量站，元数据将识别指定该站的原始数据部分、数据采样周期以及数据的准确性。目前，我们没有规范这两个组件的标准方法。

第三个问题是，在科学调查中，重要的不仅仅是数据。例如，当物理学家查看用于证明或反驳全球变暖的数据时，他们会使用计算机程序进行分析，并在这些程序中嵌入执行某些统计过程的代码。开发完全没有错误的计算机程序是非常困难的。如果其他科学家想要审核他们同事的工作或推进其工作，那么他们就需要访问程序代码。

嵌入式计算机的出现——嵌入测量仪器和射电望远镜等观测仪器——对科学造成了破坏。显然，需要更多地关注数据的归档。澳大利亚国家数据服务局等一些举措可以解决这一问题，但还需要更多。

长尾效应

2007年，克里斯·安德森出版了《长尾》一书。《连线》杂志的工作人员安德森指出，如果你查看书籍等数字物品的销售情况，你会看到如图7所示的图表。横轴显示书籍，纵轴显示销售额。你所看到的是一条标准曲线，适用于具有许多不同实例的项目，例如书籍、DVD和计算机程序。

92

该图表显示的是，在图表的最前面有相对较少数量的销量较大的书籍，而销量较小的书籍则有一条长长的尾巴。

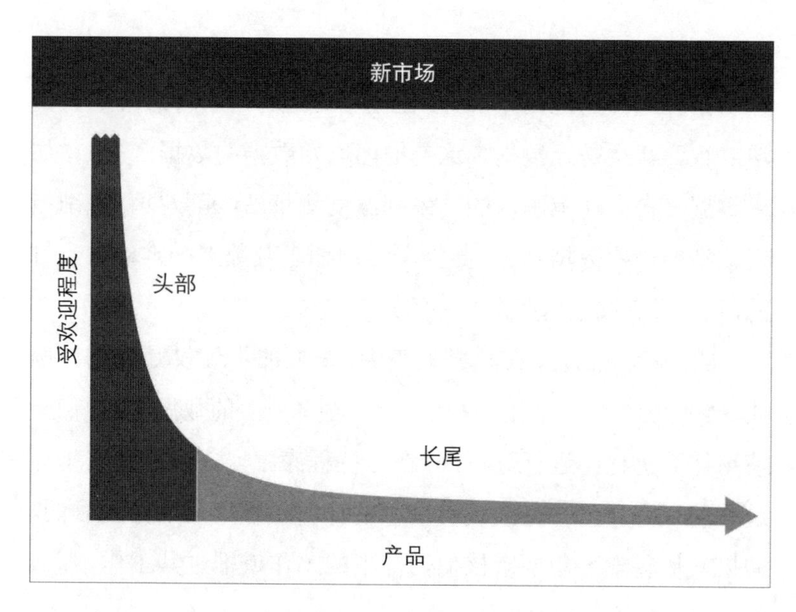

图6　长尾效应

安德森论文背后的核心思想是，数字化向数字产品的销售者承诺，在尾部销售产品的利润与在头部销售产品的利润一样高。在过去，书商因高昂的库存成本和店铺的物理限制而在销售上受到阻碍。出版商也有类似的限制：每隔几年，我就会收到一封来自我的出版商的礼貌信函，告知我的一本书即将绝版，问我是否想购买一些剩余的库存。绝版实际上意味着传统书店将不再提供更多的副本，出版商的仓库也将清空。

93

数字对象的一个关键特性是，你只需要很少的空间来存储它们：只需要一台计算机和一些大的磁盘存储器就可以保存主要出版商印刷的所有书籍。安德森的论点是，正因为如此，在长尾的末端，对于销售商和出版商来说，存在着巨大的机会，其中深奥书籍的单品销量可能很小，但总体销量可能与尾部的畅销书相当。

安德森的想法受到了许多研究人员的挑战，他们指出，他对长尾收入与头部收入相当的想法的热情可能并不成立。然而，他们错过的是，数字技术可能意味着一些深奥的书籍，比如《20世纪初的罗马尼亚蒸汽火车》，或者一张以僵尸为主角的早期日本恐怖电影DVD，可能无法与最新的紧身胸衣开膛手的销量相媲美，但它们确实与许多其他深奥的书籍一样，提供了增加销量的机会。对于主要出版商和图书连锁店来说，计算机的破坏性是以一种积极的方式；而对于采购稀有、非漫画作品的专业书商来说，破坏性则是在另一个方向。

书籍和电子阅读器

当我去吃午饭时遇到一位同事，他手里拿着一个小型的电脑设备。这是一个电子阅读器。电子阅读器允许他从出版商的网站或书商的网站下载书籍。我在2005年看到了一个电子阅读器的原型，但没有给我留下深刻印象，主要是因为屏幕的质量很差。我同事的电子阅读器要先进得多。电子阅读器的出现反映了已在音乐领域发生的一种趋势，即乙烯基唱片被CD取代，CD正稳步地被诸如iPod播放的声音文件所取代。关于电子书和 94 电子阅读器的未来正在进行有趣的辩论，其结果将决定十年或二十年后，我们是否会看到传统书店的终结。

一方面，有人指出，许多人买书是为了一种准审美体验，虽然阅读显然是买书的主要目的，但一本制作精良的艺术书，例如，有光泽的书页和高清晰度的书，是一种易于处理和观看的人工制品。也有人指出，在线图书和印刷版之间的价格差异并没有那么大；这是因为一本书的大部分成本来自它的营销、编辑、书店利润和出版商利润，而且由于这种微小的价格差异，印刷媒体不会向电子媒体转变。

然而，也有人将电子书的增长与音乐公司的增长进行了比较，他们证明在线音乐和CD之间存在相同的差异。他们还指出，文本是一种静态媒介，以视频和音频片段以及网站链接为特色的电子书，将提供更丰富的阅读体验——这将促使读者转向电子书。

现在下结论还为时过早——然而，有迹象表明电子书的增长正在加速。美国出版商协会最近与国际数字出版论坛联合开展的一项调查显示，该数字呈指数级增长。例如，2008年第三季度电子书的批发收入为1400万美元，而2009年同期的批发收入约为4700万美元。我不想指出这些数字表明n年内电子书将占据主导地位；毕竟，传统书籍的销量远远超过电子书。例如，美国出版商协会报告，传统图书的销售额为12.6亿美元。然而，数据显示出现了这种趋势，适用于iPod等设备的音乐文件越来越多。 95

2009年年末，英国书店和出版商的贸易杂志《书商》周刊发布了一项关于电子书商业可能性的调查结果。他们非常惊讶：尽管在千余名受访者中，88%的人表示他们认为电子阅读器对他们的业务构成了威胁，但许多人认为存在商业机会。一个典型的反应是，

> 通过让阅读变得更简单、更易得，以及扩大对年轻人（即移动观众）的吸引力，每个人都会受益。商业街书店需要成为

读者服务提供商——技术、一些印刷书籍（如儿童书籍、地图、艺术书籍）、建议、作者阅读、研讨会、学习中心、活动主持人等。

一种情况是，传统的连锁书店在网络竞争的压力下消失了，但当地的书店通过成为一个场所和社交中心，或许作为一个微型出版商而卷土重来。

我的观点是，目前尚不清楚传统图书销售和电子书销售之间的平衡会发生什么。然而，这里有一种情况，即具有个人观点的虚构类和非虚构类之外的电子书将会增加。例如，电子书的主要领域是百科全书和字典。维基百科在线项目已经展示了这一点：每个条目不仅包含维基百科文章上的文本，还交叉引用其他文章——通常太多——并包含许多指向相关项目的交叉链接，如论文、新闻文章和博客。

电子书的另一个发展方向是旅行书籍，在这些书中，除了传统文本外，还包含视频剪辑和照片等内容。作为一个实验，我输入了“奥赛博物馆”这个关键词，它是巴黎一家很棒的艺术画廊，由一个火车站改建而成，里面有一些19世纪和20世纪最伟大的绘画作品。我注意到有4.7万的点击量。我看了前200张照片，除了少数例外，照片的质量都非常好，很好地说明了我最喜欢的一家画廊藏品的多样性和重要性。我还查看了视频托管网站“优兔”，收到了1800个视频片段。 96

另一个可能出现爆炸性增长的领域是说明书，例如烹饪手册和汽车维修手册。在这里，展示更复杂烹饪技术的视频剪辑将穿插一些传统的文字和照片。

电子书技术还可以改变许多其他类别的书籍。例如，学校教科书可以包含解释难点的视频播客。然而，归根结底，电子书需要解决一个巨大的问题：小说和非虚构类书籍，它们都有自己的观点，比如一位作家写的第二次世界大战的历史。

有许多因素可能加速这一趋势。首先是免费材料的可得性——如果你愿意，可以使用开源文本。古腾堡项目是一个有着长久历史的项目：它始于20世纪70年代初，当时的创始人迈克尔·哈特在伊利诺伊大学的大型计算机上输入了《独立宣言》。从那时起，志愿者们在古腾堡项目的网站上输入并数字化了三万多本书。所有的书都没有版权，而且含有大量的秘籍。然而，狄更斯、福斯特和斯科特·菲茨杰拉德等作家的许多小说，都可以合法地免费下载到电子阅读器中。

谷歌可能会影响小说和非虚构作品的发展。在过去的六年中，该公司一直在以工业规模对书籍进行数字化。这项工作是与世界各地的图书馆合作完成的，比如哥伦比亚大学和康奈尔大学的图书馆。2009年，目前谷歌数字化图书的数量为1000万册。这些书中的大部分都可以部分搜索，大约100万本可以全文阅读。谷歌图书的普及，出版商现在乐于生产纸质图书的电子版本，以及产生新一代电子阅读器的技术进步，对纸质图书构成了重大威胁。

另一个可能降低图书成本的因素是在线广告。每年我都去牛津文学节——通常是和我妻子一起去。去年我们参加了一次会议，会上采访了我最喜欢的两位作者，并回答了一些问题。他们分别是唐娜·利昂和凯特·阿特金森，前者写以威尼斯为背景的犯罪小说，后者写高质量的、不可归类的小说。在我们参加他们的会议之前，我向妻子预言，绝大多数观众将是女性和中产阶层。粗略的人头统计证实，大约95%的人确实是女性——我怀疑同样比例的人是中产阶层。这次会议是营销人员的梦。

营销人员的一个关键工具是目标市场选择：了解电视节目、报纸或电视频道等媒体项目的潜在受众的人口统计信息。大多数目标市场选择都是广泛的：英国第四频道的观众有很多广泛的特点，就像《每日电讯报》的读者群，你会发现针对这些特点的广告。

与某一特定报纸的读者相比，书籍的购买者是一个更具针对性的目标市场。购买一本关于自然世界图书的人，可能会对与已购图书类似的其他图书感兴趣（亚马逊利用这一技巧向其网站的访问者展示可能的图书），他们可能会对涉及观察野生动物等活动的假期感兴趣，或者他们可能只需要一副新的双筒望远镜。 98

亚马逊销售的电子阅读器已经内置了互联网接入功能，苹果公司开发的多功能设备iPad也是如此。很快，所有电子阅读器都将具备这种功能。对于那些想购买在屏幕上打广告的图书的人来说，电子书价格可能会被压低，因为广告收入会降低电子书的成本。 99

第七章

云计算机

引　言

每当我和妻子去当地的特易购超市购物时，我们都会有一种默契：我们购买食品杂货，并在交出信用卡之前，会把特易购会员卡交给收银台的人。刷卡后，收银台上的计算机会将卡号和我们购买的商品发送到中央计算机。这种卡片的使用彻底改变了市场营销。在使用之前，结账计算机上的信息是聚合信息，例如，一周内购买了多少特定商品，以及一年中特定时间对特定商品的需求是否增加。通过将一系列购买商品与特定客户或家庭客户群联系起来，超市能够进行一些非常精明的营销。

特易购会知道：我的家人会买很多新鲜食物，如果是星期三，我们会买一些预先煮熟的食物，红酒是我们家的日常饮品。我对它所掌握的数据不会感到紧张：我获得了一个好处，因为它使得特易购能够向我发送他们认为我喜欢的并会更频繁购买的商品的特价优惠券；它还增加了特易购会员卡附带的航空里程旅行积分；特易购在了解我的家庭消费习惯方面占有优势。但我不准备与一家药品零售商达成这种交易，因为那里的购买行 100 为与健康有关。

积分卡是数据爆炸的第一个表现；互联网引发了这场爆炸，并让访问浏览器的公司和个人都可以使用它。本章的目的是看看互联网是如何将计算机转变到如此地步，以至当你把网络称为一台巨大的计算机时，你也不会被指责为科幻小说式的愚蠢。

开放数据和封闭数据

特易购的数据是一个封闭数据的例子：唯一能看到它的人是该公司雇用的营销人员。互联网用户生成了大量的封闭数据。亚马逊网站收集的数据就是一个很好的例子。每当我登录那个网站，都会看到一个首页，上面显示亚马逊认为我喜欢的书籍。展示的图书是根据我过去的购买习惯制作的。例如，如果有一周我买了一本关于合约桥牌的书，那么网站会告诉我关于这个纸牌游戏的其他书。

这些封闭数据对那些通过向公众出售或租用物品赚钱的大公司来说很有价值。“网飞奖”就是一个令人震惊的例子，说明了公司对如何改进这种数据的处理所做的努力。网飞是一家向客户出租DVD和流式传输视频的公司。它有一个客户偏好和品味的大型数据库，它能够预测，如果客户选择了特定的DVD，那么他们很有可能会想借类似的DVD。该公司使用计算机程序进行预测；然而，人们认为该程序可以改进，网飞所做的是向编程社区提出一个挑战，让他们想出一个更好的程序，奖金为100万美元。2009年，该奖项由程序员团队获得，其中大多数为美国 101

和欧洲系统公司工作。

除了封闭数据之外，还有大量的开放数据，任何可以使用浏览器的人都可以访问这些数据。2000年，加拿大黄金公司将其顶尖部门的业务放在了互联网上，让每个人都能看到。

该公司经历了糟糕的时期：其目前持有的资产似乎已被开采殆尽，黄金价格下跌，公司在一场又一场危机中蹒跚而行。黄金公司首席执行官罗伯·麦克尤恩决定将公司的勘探数据——其顶尖部门的业务——公开，并举办一场竞赛，让公众都可以处理数据，并建议未来的勘探地点。黄金公司为最佳参赛者提供了57.5万美元的奖金。

目前尚不清楚麦克尤恩激进行动的动机是什么：是绝望，还是对如何利用群众力量的洞察。不管竞赛取得巨大成功的原因是什么，各种各样的人都参与了进来：军人、学者、在职地质学家、退休地质学家和业余地质学家。结果是发现了新的、生产力很高的地点。这场竞赛使黄金公司成为黄金勘探市场的主要参与者，其价值从一亿美元增加到数十亿美元。

黄金公司的竞赛标志着向公众开放数据的爆炸式增长；不仅可以从处理公共网页中获得数据，还可以获得商业企业的计算机网络中通常隐藏的数据。

公共数据的一个例子是在线百科全书“维基百科”。这已经成为互联网的主要成功之一；它包含超过250万个条目，全部由互联网用户贡献，是我将在本章中描述的协作精神的一个例子。关于维基百科，人们通常不知道的是，它是关注自然语言处理的研究人员和商业开发人员的主要资源，例如，公司制作的程序将数千个单词总结成易于理解的形式，并在互联网上扫描文本，如涉及特定主题的报纸文章、报告、杂志文章和研究论文。

自然语言文本的一个问题是歧义。这方面的经典例子，在许多关于这一主题的教科书中被引用，比如下面这个句子

They are flying planes.

这可能有很多含义。它既可以指商用飞机的飞行员，也可以指武装部队成员驾驶的飞机的飞行员。它也可以被某人用来指天空中的许多飞机。它甚至可以用来描述儿童正在玩飞机模型。

这句话本身很难理解。然而，让它可以理解的是它的背景。例如，如果它出现在英国皇家空军招募手册中，那么它可能具有第一个含义。研究人员发现，维基百科为理解文本等任务提供了非常有用的资源。为了理解这一点，请查看以下百科全书摘录。

> 公共云或外部云描述了传统主流意义上的云计算，即通过网络应用程序或网络服务，在互联网上以细粒度自助服务方式动态调配资源，由非现场第三方提供商提供，该提供商以细粒度效用计算方式共享资源和账单。

这个摘录来自云计算的一个条目，云计算是本章的大部分内容。该摘录中带下画线的术语是对其他维基百科条目的引用。正是这些为理解一个句子提供了语境。例如，如果一篇文章中的一句话包含“云”一词，而维基百科关于云计算的文章中的许多交叉引用都出现在这篇文章中，那么你能够将其识别为一个关于特定类别的计算，而不是气象学。

另一个开放数据的例子是推特生成的数据。这是一项技术，它允许用户将短信息存放在其他人可以阅读的网站上；这有点像你通过手机收到的信息，唯一的区别是这些信息（推特）是公开的。

美国地质调查局正在利用推特来获取公众对地震的反应。它们提供关于地震严重程度的近乎即时的反馈，并使应急服务的组织速度比通过常规监测要快一点。

美国地质调查局持续收集地理代码（由第三代手机等移动设备提供的识别代码），并存储与代码相关的推文。当美国国家地震台网监测到地震时，系统会检查事件发生后的信息是否显著增加，以及信息的性质。然后，工作人员可以检查这些信息，看看地震的影响是什么。

应用程序编程接口

所以，我给你们举了一些数据库的例子，这些数据库可以由计算机出于多种目的进行操作。重要的问题是：如何获得这些数据？关键是应用程序编程接口（通常缩写为API）。

104

API是一个软件工具的集合，它使得你可以访问基于互联网的资源。例如，亚马逊有一个API，允许程序员查询亚马逊销售产品的相关数据。例如，你可以检索一本书的唯一国际标准图书编号，以及其他可用于商业活动的产品数据。这似乎是一件无害的事情，但在本章后面，我将描述这是多么具有颠覆性。

有一个名为“可编程网络”的网站，其中包含可用于访问网络资源的所有可用API的详细信息。在撰写本章时，它列出了1900多个API。

那么，你能用这些工具做什么呢？一个很好的例子是，有人将自己定位为专业书商，例如销售关于法国的书籍。亚马逊有一个合作项目，为任何在自己的网站上为公司库存书籍做广告的人提供便利。这样一个网站将包含指向亚马逊网站图书条目的链接，当法国图书网站的访问者点击其中一个链接并购买图书时，亚马逊将向网站所有者支付佣金。

一个质量好的网站，例如，有关法国的文章、法国食物的食谱和最新的新闻，将吸引那些愿意通过使用网站链接来购买书籍以支持该网站的访问者——毕竟，这本书的价格与亚马逊网站上显示的价格相同。网站上关于每本书的详细信息可以通过使用亚马逊免费提供的API获取。

这种形式的合作项目——有时被称为附属项目——现在非常流行。"合作项目"网站列出了数千个这样的项目。它包括媒体部门、食品部门以及体育和娱乐部门的组织。

API的另一个用途是通过"混搭"。"混搭"一词来自酿造业，用于描述酒精饮料（通常是啤酒）的成分组合。在计算机方面，"混搭"是使用许多API来创建混合应用程序。

第一个"混搭"是由软件开发人员保罗·拉德马赫开发的。谷歌存储了一系列地图，任何人都可以查看这些地图，并为此开发了一个API。拉德马赫开发了一个程序，在热门网站"克雷格列表"上发布房屋广告，然后在谷歌发布的地图上显示待售公寓或房屋的位置。

还有许多其他的例子。"可编程网络"网站列出了数千个示例，其中一半以上涉及映射。例如它列出了，一个为十个美国大城市提供旅游指南的"混搭"，链接到"优兔"网站上显示地标

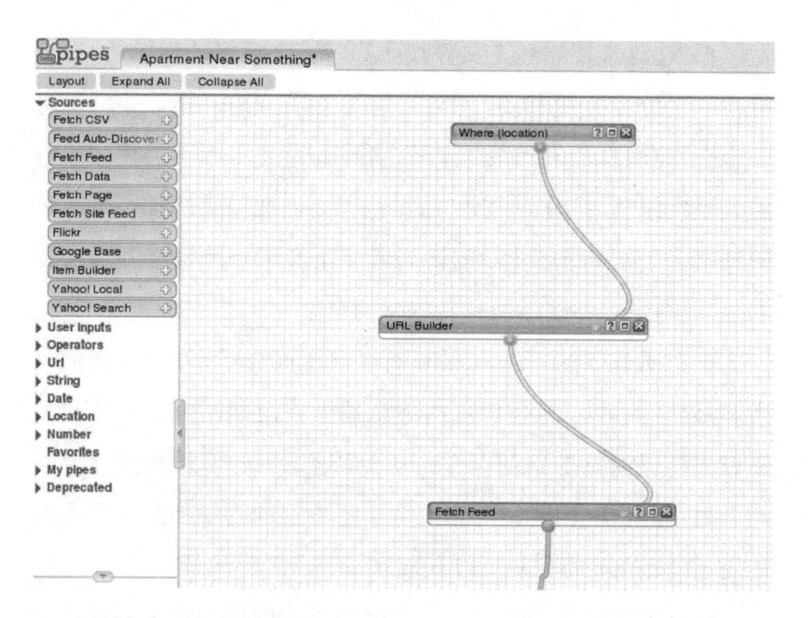

图8 “雅虎管道”项目

106 和旅游景点的视频，还有一种“混搭”允许访问者分享他们最珍爱的书籍，以及将在美国城市犯罪的地点与犯罪发生的地理位置联系起来的各种“混搭”。

开发“混搭”曾经是软件开发人员所熟悉的事情。然而，雅虎公司开发了一种技术，让任何人都可以开发“混搭”。这项被称为“雅虎管道”的技术基于一种非常简单的图形编程语言。它处理来自网络站点的数据馈送，以创建新的应用程序。

图8显示了管道程序的一部分示例。要开发管道程序，你所要做的就是将各种执行功能的框连接在一起，例如过滤数据，识别重要数据，将其传递给其他框，以及在网页上显示数据。虽然开发管道程序并不简单，但它比许多“混搭”技术要容易得多，并且为任何想要开发网络应用程序的人降低了技术障碍。

云上的服务器

本章和第一章中描述的许多趋势表明，未来十年的计算将高度集中化，许多由计算机（如普通个人电脑）执行的功能，将转移到驻留在虚拟云中的大量服务器计算机上。过去的两年中，云中的设施有所增加。

在20世纪90年代，如果你想使用软件包，例如管理公司员工的人力资源软件包，那么你会购买这样的软件包并将其安装在公司拥有的计算机上。该软件包通常通过CD或DVD等磁性介质发送给你，然后由你公司的技术人员安装在任何有需求的计算机上。然后，技术人员将负责软件的维护：安装新版本，调整软件以获得最佳性能，并回答用户提出的任何技术问题。

互联网接入速度的提高，意味着一种不同的商业模式正在出现。在这里，软件包的用户不会在自己的计算机上访问它，而是使用互联网连接访问软件，软件驻留在由软件包开发人员操作的计算机上。这样做有很多好处。

第一个优点是，它将需要更少的技术人员，或者至少使员工能够集中精力完成其他任务，而不是花时间接受培训并掌握维护软件包所需的最新知识。

第二个优点是，它减少了维护软件版本所需的工作量。大型软件系统经常会更新，例如，当足够多的客户需要新功能时，或者当法律发生变化，要求公司采取不同的行为时。对于不具备技术技能的员工来说，安装更新的过程可能是一场技术噩梦。由于只有一个软件副本驻留在一台计算机上，更新过程更加简单。

第三个优点是降低了计算机硬件成本。云计算概念设想软件和它所需的数据都驻留在云中。理论上，一家公司在云中访问一个软件包所需的一切只是一台非常精简的计算机，几乎不需要文件存储，实际上就是20世纪70年代所谓的“哑终端”。

108 云计算的支持者声称：它使公司的业务更加灵活，降低了人力和硬件成本，使用户能够在任何地方连接到软件，而不仅仅是从办公环境中的计算机，实现了更高的硬件效率，因为软件包所在的服务器可以更有效地使用我在第一章中详细介绍的闲置资源，更高效地使用资源，并满足绿色议程的要求。它还可以实现更高程度的安全性，因为使用的所有数据都保存在数量有限的服务器上，而不是分散在大量个人计算机上，这些计算机容易受到技术安全攻击和物理安全问题引起的攻击。

云计算也存在潜在的问题。首先，当一家公司破产时会发生什么？如果没有一家公司热衷于接管破产的公司，这将使已倒闭公司的客户陷入非常尴尬的境地。其次，尽管有一种观点认为，使用云计算可以增强安全性，但存在一个问题，即如果云公司的安装过程中发生安全违规，则影响会更大。如果一家公司的数据被盗，那么这只是该公司的一个主要问题：如果数据从一家云公司被盗，其所有客户都将受到威胁。

云计算不仅仅局限于工业和商业软件包，它开始影响家庭计算机用户。电子邮件就是一个很好的例子。许多计算机用户已经转向基于智能手机的电子邮件程序和基于网络的邮件系统。谷歌应用程序就是一个例子。这是一套基于微软办公软件的工具，包括文字处理器、电子邮件程序和日历。如果你对基本功能满意，那么这些应用程序是免费的。不仅应用程序是免费的，而且你使用的存储也是免费的，例如用于字处理文档的存储。

尼古拉斯·卡尔是计算机进化方面最有趣的作家之一。他写了两本书，这两本书将信息技术社区划分开来。第一本书是《IT不再重要》。在这本书里，他认为，随着信息技术变得更加普遍、标准化和廉价，信息技术在商业中的战略重要性已经减弱。在这篇文章和其他一些学术性文章中，他在2004年指出，随着硬件成本的下降，公司会购买他们的计算业务，就像购买电力和天然气等公用事业一样。云计算就是一个例子。 109

他的第二本书《大转换：重连世界，从爱迪生到Google》，进一步探讨了效用计算的主题，并用电力行业增长的类比来揭示计算领域正在发生的事情。卡尔指出，发电行业一开始几乎就是一个家庭手工业。一家需要电力的公司会安装自己的发电机，并由其工程人员进行维护；这导致了对具有重型电气工程技能的工程师的大量需求。他指出，像爱迪生这样的先驱改变了发电的世界，使电力由大型公用事业公司进行生产和分配。

在一系列巧妙的类比中，卡尔展示了计算中同样的力量在发挥作用：服务正逐渐迁移到驻留在互联网上而不是某些公司场所的大型服务器集合。我将在最后一章探讨其影响，包括：控制权从媒体和其他机构转移到个人，对安全和隐私的主要担忧，以及高技能知识工人的工作输出。

社交计算机

1995年，一个网站在互联网上建立了。这个名为“维基维基网”的网站改变了我们使用计算机的方式。它的开发是为了帮助软件工程领域的程序员有效地互相交流，它是白板的电子版。工程师可以在白板上添加想法，删除他们认为不相关的任 110

何内容，并修改其他人的想法。

在该网站上线之前，从计算机到互联网的流量主要是单向的。如果你想查看网页，可以单击链接，将几个字节的短信息发送到网络服务器，包含数千字节的页面就会返回并显示在浏览器上。“维基维基网”开始了一个过程，计算机用户和互联网之间的互动变得更多是双向的。

早期有一些双向沟通的例子。例如，在线书店亚马逊邀请其网站的用户对他们阅读的书籍发表评论。这些评论随后被插入到包含该书详细信息的网页中。然而，这种交互并不是即时的：它被人工编辑器截取以进行检查，然后插入。

“维基维基网”的发明者是沃德·坎宁安，一位研究人员和顾问。他希望能够与同事交流，让他们可以使用虚拟白板。坎宁安的想法催生了一个被称为“维基”的软件系统。维基所做的是在互联网中维护一个文档或一系列文档。然后，维基用户可以修改其内容，甚至可以删除一些内容。大多数与计算机相关的发明，都可以被视为由软件或硬件的某些改进推动的。维基的发明非同寻常，因为它在概念上很简单，只需对现有系统进行很少的更改。

维基百科可以说是维基最好的例子。这部在线百科全书是由数千名志愿者开发的；著名的科学杂志《自然》进行了一项有争议的研究，将与之同样享有盛誉的《大英百科全书》进行了比较，得出的结论是，两者在准确性方面接近。

111

维基有很多应用程序。例如，公司使用它们与客户互动：获取对当前产品和服务的看法，并发现客户想要什么。另一个例子是使用维基来运行大型项目，尤其是那些分散在多个地理位置的项目。我自己的大学——一所绝大多数学生在家学习的大学——使用维基让学生能够相互协作；它提供了一种虚拟的互动形式，类似于传统大学中的社会互动。

维基预示着一个双向交流的时代，现在这种交流已经成为互联网上的常态。社交网站“脸书”就是一个很好的例子。“脸书”是一个允许任何人加入的网站，可以建立自己的个人页面，也可以与被称为“朋友”的其他互联网用户群进行交流。“脸书”提供的众多工具之一是“墙”。这是一个共享区域，用户和他们的朋友可以在其中编写他们都可以访问和阅读的信息。

维基仅仅是越来越多地使用计算机进行社交和开发协作产品的一个组成部分。“推特”是一个允许任何连接到互联网的人发送短信息（称为“推文”）的网站，无论是移动的还是其他的。“网络相簿”网站允许用户在网上存储照片，任何有电脑和网络连接的人都可以查看和评论照片。“优兔”是一个视频共享网站，允许用户在互联网上发布短视频，并且像“网络相簿”一样，允许任何人对其发表评论；“美味”是一个允许计算机用户共享收藏夹的网站。“掘客”是一个允许用户共享世界各地链接和故事的网站。“掘客”的用户可以对这些故事进行投票，这些故事将被提升到网站上更突出的位置（一个被称为“挖掘”的过程）或降低显著性（一个被称为“掩埋”的过程）。

产品的协同开发也是过去十年中蓬勃发展的一个领域。维基百科又是一个很好的例子：大量志愿者开发了一部非常全面的百科全书。然而，还有其他一些例子。

一个学术性的例子是“开放湿件”。这是麻省理工学院的一个项目，旨在分享生物学知识，并通过多个合作机构存储研究

112

成果，如文章和实验方案。

计算机可作为数据挖掘工具

计算机与互联网之间的连接和存储技术的进步带来的后果之一，是生成、存储并使计算机用户易于访问的大量数据。通常，存储的数据量越大，可以从中获取的信息就越多，但处理成本往往更高。

史蒂文·莱维特和史蒂芬·杜布纳在其新书《超级金融经济学》中描述了一个最近的数据挖掘示例。他们描述了如何挖掘潜在恐怖分子的财务记录和习惯以获取信息；这包括消极特征和积极特征。消极特征的一个例子是，他们不太可能有储蓄账户；积极特征是，他们每月的存款或取款时间不一致。

大规模数据使用最生动的例子之一，是语言计算机翻译的最新进展。自20世纪60年代以来，使用计算机在一种语言和另一种语言之间进行翻译一直是研究人员的目标。将A语言翻译成B语言的通常方法是制定一套规则，描述A语言的句子结构，处理A语言中需要翻译的文本，使用规则识别其结构，然后使用B语言的句子结构规则将文本转换为该语言中的对等文本。

113

在所谓的“机器翻译”方面的进展是稳定的，但很缓慢。然而，谷歌最近的工作加快了这一进程。该公司的研究人员开发了一种新技术，该技术依赖于使用大量现有文本。研究人员使用了一种被称为“机器学习”的技术来进行翻译。机器学习是一种向计算机程序提供大量案例和结果，并试图从所提供的数据中推断出一些假设的方法。在谷歌的案例中，机器学习计划获得了2000亿字的文本以及联合国提供的其他语言文本；然后，它学习了从一种语言到另一种语言的句子转换方式，而不知道所使用的任何语言的结构。与使用语言规则描述的程序相比，未来几年有望有所改进。

兰德尔·斯特罗斯在其著作《谷歌星球》中描述了一个阿拉伯语句子如何被商业翻译系统转换成了“阿尔卑斯山白色新存在磁带，注册用于咖啡确认拉登”，而谷歌系统几乎完美地将其翻译成“白宫确认存在新的本·拉登磁带”。

在本章的开头，我描述了积分卡如何为特易购等零售商提供有用的信息。这是开发可用于各种营销目的的数据库并使用计算机对其进行处理的示例。示例还有很多。数据挖掘的一个主要应用是市场篮子分析。在这里，公司将保留过去的采购和其他数据，如人口统计信息，以提高其销售。例如，亚马逊公司通过使用过去的销售来告知客户他们可能感兴趣的商品，这些商品与他们过去的销售有某种关联；例如，购买犯罪小说的客户再次登录亚马逊网站时，几乎总是会看到新出版的犯罪小说。

114

使用市场篮子分析的另一个例子是识别所谓的“阿尔法客户”。这些客户的职业生涯让他们面对大量的人。他们可能是经常在电视上看到的名人，也可能是在报纸上看到的名人，也可能是在研讨会和股东大会上被聆听的商界领袖。重要的是，这些人的生活方式经常被模仿，或者其建议被接受。

这种数据挖掘的应用只有通过一种被称为病毒式营销的概念才能盈利，即“脸书”等社交网站放大消费者之间的对话并传播这些对话。有许多例子可以说明这是多么成功。在互联网出现之前，电影《布莱尔女巫计划》不只是一部邪典电影，而且是一部小众的邪典电影；网络口碑使它获得了巨大的成功。

聚合数据技术的一个有趣商业用途是预测市场。这在许多方面与投资者买卖公司股票的股票市场相似。然而在预测市场中，它是一项预测：它可能是关于谁将获得奥斯卡奖，政党的选举多数是什么，或是诸如汇率重大变化等经济事件的预测。当此类市场流行时，它们可以进行高度准确的预测。例如，一个涉及奥斯卡预测买卖的市场准确预测了2006年奥斯卡36大类提名中的32项，以及8大类获奖者中的7项。

115

从互联网现象来看，预测市场现在已经成为大公司用于经济预测的东西。例如，谷歌等公司利用内部预测市场来确定业务政策。

在本书的开头，我将计算机定义为：

> 计算机包含一个或多个处理数据的处理器。处理器连接到数据存储器。操作者的意图是通过若干输入设备传达给计算机的。处理器执行的任何计算结果都将展示在若干显示设备上。

你可能会认为这就是我的学术性：依赖定义，厘清所有问题，并试图语义清晰。然而，这一定义的要点是，它不仅涵盖了你办公桌上用于文字处理和网络浏览等任务的计算机、帮助飞行的计算机或嵌入iPod的计算机芯片，还包括由数亿台计算机组成的、嵌入云的更大计算机；我希望这一章能让你相信这一点。

116

第八章

下一代计算机

引　言

自国际商用机器公司开发出第一台大型机以来，计算机的基本体系结构已经保持了六十年不变。它由一个处理器组成，该处理器可以逐个读取软件指令并执行它们。每一条指令都会导致数据被处理，例如通过将数据添加到一起；数据存储在计算机的主存储器中，或存储在某种文件存储介质上；或者被发送到互联网或其他计算机。这就是所谓的冯·诺伊曼体系结构；它是以约翰·冯·诺伊曼的名字命名的，他是一位归化的美国数学家，在20世纪40年代和50年代处于计算机研究的前沿。他的关键思想至今仍占据主导地位，那就是在计算机中，数据和程序都存储在同一地址空间的计算机存储器中。

冯·诺伊曼体系结构几乎没受到什么挑战。在最后一章中，我将介绍两种未来的计算方法，它们涉及完全不同的体系结构。我还将研究一个被称为“神经网络体系结构”的子架构，它与大脑中的神经结构有一些相似之处。但是，一段奇怪的历史 117 可能会重演。

函数式编程和非标准体系结构

20世纪80年代，计算机研究人员的研究经费大幅增加，英国、欧洲大陆和美国的资金来源多得令人难以置信。如果认为这是因为政府意识到了计算机的潜力，那就太好了。然而，它的出现是因为恐惧：对日本工业的恐惧。日本通产省宣布了一项向计算机技术提供资金的重大计划，西方国家政府在经历了日本电子和汽车行业所造成的巨大破坏后，担心其仍在发展的计算机行业也会发生类似的事情。

英国设立了阿尔维项目。这是政府部门和科学研究委员会之间的一项联合计划，旨在推动研究和提供训练有素的信息技术人员。这项研究有许多线索，其中之一就是新颖的计算机体系结构。

在20世纪80年代，许多研究人员描述了软件开发和冯·诺伊曼体系结构的问题。一些人指出，用传统编程语言开发的软件常常出错，而且计算机读取数据、处理数据和写入存储的艰苦过程过于详细和复杂。其他研究人员指出，在未来的几十年，硬件技术将发展到可以在单个芯片上嵌入大量处理器的地步：这在计算能力方面提供了重大机会，但由于程序员试图在处理器之间共享工作，也会产生更多错误。

为了解决这些问题，计算机科学家开始开发一种新的编程语言，被称为“函数语言”。这些语言没有常规编程语言的分步 118 功能；相反，它们由一系列数学方程组成，这些方程定义了计算

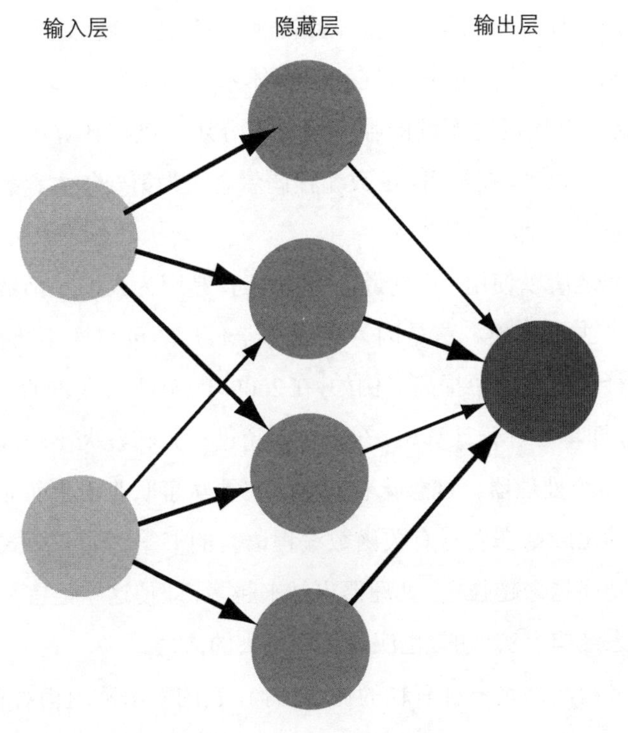

图9　一个简单的神经网络

的最终结果。由计算机“解”这些方程并给出正确的结果。

许多函数语言被开发出来，最著名的四种是FP、Haskell、ML和Miranda。还有许多执行这些语言程序的计算机包被开发出来。不幸的是，出现了一个问题：程序驻留在冯·诺伊曼计算机上，这导致执行速度非常慢，因为传统的体系结构无法胜任这项工作。

119

由于传统体系结构的问题，许多研究机构利用硬件处理器阵列创建了新型计算机。这些处理器将采用功能程序，并处理程序的内部表示，将其简化为可以高效执行的形式。

由于计算机科学家进入网格计算等其他领域，这方面的结果并不乐观，函数编程研究和新型体系结构也在下降。直到最近，这一有趣研究领域的唯一剩余部分还可以在计算机科学学位课程中找到，在那里，函数编程偶尔会成为第一年本科教学的一个基本要素。

我之所以使用“直到最近”这个词，是因为人们对函数语言重新产生了兴趣。这是因为多处理器芯片的可得性不断提高，其运行速度也大幅提高。例如，在2009年，英特尔宣布了一款处理器，即单芯片云计算机，在一块不超过一张邮票大小的硅晶片上有48个处理器。这意味着，例如，主要从事职业销售而非学术销售的出版商在发行有关函数编程语言的书籍。随着台式计算机的功能越来越强大，处理器也越来越多，即使这些语言是在传统体系结构上实现的，也应该实现更大的渗透。

神经网络是一种有趣的体系结构，它的应用范围相对有限。这是一种经常在冯·诺伊曼计算机上模拟的体系结构。它用于模式识别应用和预测短期趋势。

120

罗切斯特大学医学中心的安德里亚·道森、加利福尼亚大学旧金山分校的理查德·奥斯丁和布里格姆妇女医院的戴维·温伯格，开发了一种基于神经网络的癌细胞检测系统；该系统可以检测癌细胞，准确度与人类检测相当。

伦敦大学玛丽皇后学院的研究人员将神经网络用于安全应用中，即使入侵者可能正在移动，他们也能聚焦建筑物中可疑入侵者的面部。

萨塞克斯大学的研究人员使用了我在第四章中描述的遗传编程思想，创建了几代神经计算机，并选择了最好的神经计算机来解决追捕和逃脱领域的问题，因为某人或某物试图躲避追捕者。

那么，什么是通常被称为“神经网络”的神经计算机？最初，这种计算机的想法源于对大脑神经结构的研究。图9显示了一个简单的示意图。它由一个输入层组成，该层可以感知来自某些环境的各种信号，例如构成人脸图片的像素。在中间（隐藏层），有大量的处理元素（神经元），这些处理元素被排列成子层。最后，还有一个输出层，它提供了一个结果，例如，这可能是安全系统上的一个简单窗口，例如，当识别出航空公司乘客是潜在恐怖分子时。

这项工作是在神经计算机的中间层完成的。其发生的机制是，通过向网络提供要识别的趋势或项目的示例，对网络进行培训。训练的目的是加强或削弱中间层处理元素之间的联系，直到它们结合在一起，当向它们呈现与先前训练的示例相匹配的新案例时，它们会产生一个强信号；当遇到与示例不匹配的项目时，它们会产生一个弱信号。

神经网络是在硬件中实现的，但大多数则是通过软件来实现的，其中的中间层是在执行学习过程的代码块中实现的。

在研究其他新的体系结构之前，值得一提的是，尽管最初的动力是利用神经生物学的思想，在考虑大脑过程的基础上开发神经体系结构，但目前商业实现中使用的内部数据和软件与人脑之间几乎没有相似之处。

量子计算机

在第四章中，我讨论了超大规模集成电路技术如何使计算机公司开发出越来越强大的计算机；功率的增加是由于在单个芯片上挤压越来越多处理器的能力的增强。

冯·诺伊曼计算机的基础是使用集合成字节的二进制数字（0或1）存储数据。量子计算机则以量子位数据存储；这些也可以原子、离子、光子或电子来实现。这些量子位不仅可以作为数据的存储元素，还可以组合在一起来实现硬件处理器。

为了了解量子计算机的功能，牛津大学的学者和物理学家、该领域的先驱之一戴维·多伊奇计算出，一台普通的30量子位计算机可以在每秒10万亿次浮点运算速度下工作。这与21世纪头十年运行的超级计算机所达到的速度相当。

使用量子思想实现计算机的一个问题是，担心观察量子位可能产生的影响。例如，试图检查一个量子位的状态可能会改变它的状态。这意味着从量子计算机读取数据可能会非常困难，而这一过程在传统计算机上非常容易。

令人高兴的是，有一种被称为量子纠缠的现象，它得到了量子计算机开发人员的帮助。2006年，美国商务部国家标准与技术研究所的物理学家，在10月19日出版的科学杂志《自然》上报告称，他们朝着将纠缠转化为可用于读取量子数据的技术迈出了重要一步。他们展示了一种重排纠缠原子对的方法——一种被称为提纯的过程——其中包括缠绕的两对铍离子。这意味着可以间接观察量子计算机中的数据，而不影响其价值。

量子计算才刚刚起步。迄今为止，研究结果的规模还非常小：例如，加拿大的一家初创公司D-Wave演示了一台可工作的16量子位的量子计算机。计算机解决了一个简单的数独难题和其他模式匹配问题。现在，与传统计算机的性能相比，这个中等规模问题的解决方案在性能方面并没有太大的成就。然而，这是一个戏剧性的概念证明。

量子计算机很重要，因为一台基于量子物理学思想的成功计算机可以颠覆目前存在的许多技术。它将产生最大影响的领域之一是密码学。许多现代密码方法依赖于这样一个事实，即解决某些问题非常困难，这些问题被称为棘手问题。然而，研究人员指出，一台相对有限的量子计算机就可以解决这些问题；事实上，这个学术社群所面临的挑战之一就是如何做到这一点。 123

例如，2001年，国际商用机器公司和斯坦福大学的计算机科学家证明，量子计算机可以通过编程来查找数字的素因子（素因子是一个能整除给定正整数的数字，并且不能被进一步约除；例如，33的素因子是3和11）。研究人员使用一台7量子位的计算机来查找15（5和3）的素因子。素因子的确定是使基于互联网的密码学获得成功的问题之一。

不过，在计算方面，这并不是什么伟大的成就；例如在2005年，德国联邦信息技术安全局的研究人员使用一组传统计算机来分解一个200位的数字。然而，它代表了一个重要的概念验证，如果能将其产业化，它将威胁到我们用于发送安全数据的大部分技术。

DNA计算机

脱氧核糖核酸（DNA）是一种含有遗传指令的核酸，用于确定生物体和某些病毒的发育和生命。当生物学家提到DNA时，他们经常用计算机程序来谈论它：一个在执行的程序，控制我们的身体生长，并构成我们的基因图谱。

DNA包含由两个被称为核苷酸的简单单元长聚合物组成

的数据。在DNA中也有糖和磷酸盐。附着在每一种糖上的是四类（A、T、C和G）分子中的一类，被称为碱。这些是编码我们基因发育中所用信息的基本单位。

因此，我们就有了一台潜在的包含数据的计算机和一个处理这些数据的程序。早在1994年，美国计算机科学家伦纳德·阿德尔曼就表明，基于DNA的计算机可以解决有向汉密尔顿路径问题。这是第四章描述的旅行推销员问题的一个特殊版本。阿德尔曼的程序处理的数据点数量很少：台式计算机可以在短时间内获得正确的解决方案。然而，它代表了一种概念验证。

这些是DNA计算机为解决问题可能执行的步骤：DNA链将被视为解决问题的数据。这些将使用在基础中被找到的A、T、C和G编码进行编码。

然后，这些分子会被混合，许多DNA链会粘在一起。每一组绞线都代表了问题的可能解决方案。无效答案将通过一系列化学反应消除，剩下的将是问题的一个或多个解决方案。

在阿德尔曼完成划时代的工作后，许多研究小组意识到使用人类遗传物质的潜力，开始研究DNA和冯·诺伊曼计算机之间的类似物。罗切斯特大学的研究人员利用遗传物质开发出了被称为"逻辑门"的电子电路。正如你在第一章中所记得的，逻辑门是一种电子设备，它接受许多位，并产生另一个二进制位作为输出，该二进制位取决于输入。例如，"与"逻辑门的两个输入为1时产生1，否则产生0。正如DNA的组成部分是生命的基本组成部分一样，门也是计算机的基本组成部分；例如，它们用于实现加法和减法等算术函数。

DNA计算机显然与当前的计算机有着根本的不同，因为用于生产它们的材料是化学性质的——当前的计算机是由金属和硅制成的。然而，它们并不像量子计算机那样背离冯·诺伊曼的思想。它们所提供的是一种计算范式，这种范式提供了大量的存储和很少的功耗。就像量子计算机一样，它们只是刚刚从概念验证阶段出现。

反乌托邦视角

乔纳森·齐特林是一位杰出的互联网治理教授。他写了一本关于互联网未来的重要著作——不是技术未来，而是一个可能受到政府和工业行动制约的未来。在这本书的结尾加上一章对计算机的看法有些不合时宜，这似乎有些奇怪；毕竟，这整本书都是关于计算机的，并集中于它所取得的惊人进步——那么，为什么要以互联网作结，为什么要终结于一个悲观的音符呢？第一个问题很容易回答：这本书一直强调，你不能让计算机脱离它现在运行的网络环境。第二个问题的答案是，如果不考虑一些缺点，很容易被计算机的奇思妙想冲昏头脑。我在第三章中做了一点，在那里，我简要探讨了一些安全和隐私问题。

计算机可以解决重大问题而无须考虑背景的想法似乎仍然很流行。这方面的一个例子是与英国政府信息技术项目相关的浪费；过去十年来，国家卫生信息技术项目等大型项目出现资金渗漏。例如，纳税人联盟最近的一份报告记录了大量报告预算超支的政府项目。它详细说明了约190亿英镑的超支，其中基于信息技术的项目贡献了120亿英镑。

126

在他的书的第一部分，齐特林描述了计算机的历史。它始于个人计算机由专业人员维护，软件系统满足狭窄的要求，例如制作员工工资单的时候——这发生在20世纪60年代左右。然后，他描述了个人电脑逐渐成为许多人电子家具一部分的过程，以便非技术用户可以安装他们想要的任何软件，直到通过互联网，它成为一种大规模创造力和大规模协作的工具。

在他的书中，齐特林探讨了这种蜕变的各个方面之间的紧张关系：第三章详述的安全性下降，第五章详述的隐私下降，以及计算机和互联网作为公共财产能够提供信息、促进协作和鼓励创造力的概念。

齐特林介绍了计算机发展中可能出现的一些阻碍因素，以及未来可能使用它做些什么。一个潜在的障碍是允许访问互联网的设备，其访问方式只能由设备制造商控制。这些设备包括游戏机、MP3播放器、数字电视录像机和电子阅读器。

这些设备要么完全封闭，要么几乎完全封闭，用户只能在严格条件下访问和修改底层软件——通常这些用户为设备开发第三方软件，只有在达成商业协议后才允许访问。

齐特林暗示的一个可能未来是，在互联网中，通过家庭计算机终端进行访问，而当前个人电脑的硬件能力很低——它们只是连接到互联网，并且有一系列捆绑设备以同样的方式限制用户，例如，朝鲜政府限制其国内的外部无线接收。

127

齐特林的主题之前是尼古拉斯·卡尔在《大转换》一书中提出的想法。卡尔的重点是商业。在书中，他将互联网的增长与发电的增长进行了比较。在19世纪下半叶和20世纪初，发电几乎是作为家庭手工业进行的，每家公司都维护自己的发电机，并雇用重型电气工程师来维护发电机。卡尔的书展示了托马斯·爱迪生和塞缪尔·英萨尔如何通过开发一种基础设施，使国内用户和工业用户从电网中获取电力，从而彻底改变了发电行业。向电力的转移很快，这是发电技术进步的结果，也是用交流电取代直流电的根本步骤。

卡尔将发电量的增长与互联网的增长进行了比较。正如你在第七章中所看到的，计算机能力越来越趋向于集中化。例如，越来越多的网站安全地存储你的数据；网格允许互联网消除个人电脑的处理压力；而且，越来越多的公司通过网站提供软件功能，而不是通过提供可在个人电脑或服务器上安装软件的设施。

基础设施已经存在，可以用精简的、基于计算机的消费设备取代个人电脑，这些设备具有屏幕、键盘和足够的硬件，允许用户与互联网进行通信，并在固件中嵌入不可修改的软件。

卡尔的论文也得到了另一位作家吴修铭的支持。在《总开关》一书中，吴先生还采用了基于电视和其他传播媒体发展的历史视角。他指出，由于商业压力，那些曾经自由开放的技术最终变得集中和封闭。 128

因此，未来计算机的另一个形象是作为一个嵌入式设备嵌入消费电子产品中，通过浏览器与互联网交互；使用远程服务器上的文字处理器执行诸如开发文本等活动。这将是一件电子设备，防止用户通过软件修改设备。几十年后，我们可能会像看早期发电硬件的照片一样看我们目前使用的台式机或笔记本电脑。 129

索　引

（条目后的数字为原书页码，见本书边码）

A

Abhu Ghraib incident 阿布格莱布监狱虐囚事件 91
Active badge 电子识别卡 41
Adams, Joel 乔尔·亚当斯 58
Address 地址 32
Address bus 地址总线 32
Adobe Photoshop 奥多比图像处理软件 14, 89
Adleman, Leonard 伦纳德·阿德尔曼 72, 125
Adwords 关键词广告 83—84
Aerodynamic design 气动设计 57
AirStrip OB 远程监测孕妇生命体征的应用程序 44
Alpha customer 阿尔法消费者 115
Alvey Programme 阿尔维项目 118
Amazon 亚马逊公司 9, 22, 58, 79—80, 98—99, 101, 105, 115
Ambient informatics 环境信息学 41
Anatomy Lab 解剖实验室 44
And gate 与门 26—27, 125
Anderson, Chris 克里斯·安德森 92
Apache web server 阿帕奇网络服务器 82
Application Programming Interface 应用程序接口 104—106
Approximate algorithm 近似算法 52
Artificial intelligence 人工智能 20, 54
ASCII 美国信息交换标准代码 25
Associate programme 合作项目网站 105
Association of American Publishers 美国出版商协会 95
Asymmetric cryptography 非对称密钥加密 72
Audio podcast 音频播客 88
Austin, Richard 理查德·奥斯丁 121
Australian National Data Service 澳大利亚国家数据服务局 46, 92
Australian Square Kilometre Array 澳大利亚的平方千米阵列 45

B

Beowulf cluster 贝奥武夫机群 57—58
Binary number system 二进制数系 25
Bin packing problem 装箱问题 52
Biological computing 生物计算 22
Biometric identification 生物识别 75
Biometric smart card 生物识别智能卡 72
Bit 位 1
Blog 博客 86
Brixey, Shawn 肖恩·布里克西 91
Borders UK 鲍德斯英国公司 79
Bowen, Debra 德布拉·鲍温 42
British Telecom's Vision programme 英国电信的愿景项目 12
Broadband 宽带 10
Brom, Tim 蒂姆·布朗 58
Browser 浏览器 9

Bug 故障 28
Byte 字节 1

C

C# 微软公司发布的一种由 C 和 C++ 衍生出来的面向对象的编程语言 11
Caching 高速缓存 35
Caeser cipher 恺撒密码 70—71
Carr, Nicholas 尼古拉斯・卡尔 61—62, 84, 109—110, 128
CDC 美国数据控制公司 56
Christensen, Clayton 克莱顿・克里斯滕森 81
Cipher text 密文 70—71
Citizen journalism 公民新闻 86—87
Client computer 客户端计算机 9
Climategate “气候门”事件 47, 88
Climatic Research Unit (CRU) 英国气候研究所 46—47, 87
Closed data 封闭数据 101
Cloud computing 云计算 12, 21—22, 61, 100, 104, 107—109
Cocks, Clifford 克利福德・柯克斯 72
Computer chess 计算机国际象棋 7
Computer Economics 计算机经济学 85
Computer program 计算机程序 4, 10
Computer virus 计算机病毒 14, 21
Core memory 磁芯存储器 28
Craigslist 克雷格列表 106
Cray Corporation 克雷公司 56
Cray XM-P48 克雷巨型机 56
Cryptography 密码学 70, 123
Cunningham, Ward 沃德・坎宁安 111

D

Data.gov 美国政府的数据官网 13
data.gov.uk 英国政府的数据官网 13
Dawson, Andrea 安德里亚・道森 121
Database server 数据库服务器 59
Deep Blue 深蓝(美国商用机器公司生产的国际象棋电脑)55
Delicious “美味”网站 112
Denial of service attack 阻断服务攻击 63, 69
Deutsch, David 戴维・多伊奇 122
D-Wave 一家初创量子计算公司 123
Dial-up 拨号上网 10
Diffie, Whitfield 惠特菲尔德・迪菲 72
Digg “掘客”网站 87, 112—13
Digital camera 数码相机 14, 34
Digital signature 数字签名 73
DNA 脱氧核糖核酸 124—125
DNA computer DNA 计算机 23, 124—126
Domain Name System 域名解析系统 9, 59
Dombis, Pascal 帕斯卡・多姆比斯 91
Dubner, Stephen 史蒂芬・杜布纳 113
Dumb terminal 哑终端 108, 128
Dumpster diving 垃圾搜寻 69
Dynamic Random Access Memory (DRAM) 动态随机存取存储器 34

E

e-book 电子书 95, 97

eCray XT5 Jaguar 克雷 XT5 美洲豹超级计算机 57
Electrical relay 继电器 28
Electrically Erasable Programmable Read Only Memory (EEPROM) 电可擦可编程只读存储器 33
Electronic valves 电子阀 28
Elliot 803 computer 埃利奥特 803 型计算机 1—2, 24, 28
Ellis, James 詹姆斯・埃利斯 72
Entanglement 纠缠 123
Epocrates 一种药物词典应用程序 44
e-reader 电子阅读器 94—95, 99, 127
Erasable Programmable Read-Only Memories (EPROM) 可擦可编程只读存储器 33
Evolutionary programming 进化编程 53
Expert system 专家系统 55

F

Facebook 脸书 112, 115
Fetch-execute cycle 提取执行周期 27—28
Firewall 防火墙 21, 65, 67—68
Flash memory 闪存 33
Flickr "网络相簿" 网站 14, 112
Folding@home 一个研究蛋白质折叠、误折、聚合及由此引起的相关疾病的分布式计算项目 17, 60
Fonda, Jane 简・方达 90
Fourth paradigm 第四范式 47—49
FP 编程语言 119
Freedom of Information Act《信息自由法》47
Friv 在线游戏网站 16
Functional programming 函数编程 118
Functional programming language 函数编程语言 119

G

Genome sequencing 基因组序列 21
Genetic programming 遗传编程 53
Gigabyte 千兆字节 1
Goldcorp 加拿大黄金公司 102
Google Apps 谷歌应用程序 23, 109
Google Inc. 谷歌公司 15, 23
Gray, Jim 吉姆・格雷 47
Grid computing 网格计算 21, 60, 120

H

Halstead, Dirck 德克・霍尔斯特德 91
Hamiltonian path problem 汉密尔顿路径问题 125
Hard disk memory 硬盘存储器 4—5, 35
Hart, Michael 迈克尔・哈特 97
Haskell 一种标准化的通用纯函数编程语言 119
Hellman, Martin 马丁・赫尔曼 72
High Dynamic Range imaging 高动态范围成像 89
Home page 主页 36
Hopper, Grace 格雷丝・霍珀 28
HSBC 汇丰银行 76

Human Genome Project 人类基因组计划 17, 51

I

Intel 英特尔公司 3, 58
International Digital Publishing Forum 国际电子出版论坛 95
Internet 互联网 8, 12
Intrusion detector 入侵探测器 21
iPad 苹果平板电脑 99
iPhone 苹果手机 16, 44
iPod 苹果数字多媒体播放器 6, 12, 21, 24, 41, 85, 94—95, 116
iPot 一个能让咖啡或汤整天保持温暖的水壶 38, 45
iTunes 苹果媒体播放应用程序 80

J

Java 一种编程语言 11

K

Kasparov, Gary 格雷·卡斯帕罗夫 55
Kerry, John 约翰·克里 90
Keyboard 键盘 6
Kilobyte 千字节 1
Krums, Janis 贾尼斯·克鲁姆斯 89

L

Large Synoptic Survey Telescope 大型巡天望远镜 91
Levitt, Steven 史蒂文·莱维特 113
Logic bomb 逻辑炸弹 65
Logic gate 逻辑门 125
Linux 一种免费使用和自由传播的类UNIX操作系统 57, 82—83
Loyalty card 积分卡 101

M

Machine learning 机器学习 114
Main memory 主存储器 4—6, 26
Malware 恶意软件 64
Market basket analysis 市场篮子分析 115
Mashing 混合 22
Mashup 混搭 106—107
Mass collaboration 大规模协作 17—18
Mass computing 大规模计算 17
Maxfield, Clive 克莱夫·马克斯菲尔德 32
McEwen, Rob 罗伯·麦克伊文 102
Megabyte 兆字节 1
Melissa virus 梅丽莎病毒 63—64
Metadata 元数据 48
Microsoft Office 微软办公软件 23
Microwulf cluster 米罗沃夫集群 58
MITI 日本通产省 118
Miranda 一种编程语言 119
Mitnick, Kevin 凯文·米特尼克 76
ML 一种通用函数编程语言 119
Mobile phone 移动电话 34, 41, 43
Modem 调制解调器 26
Moore, Gordon 戈登·摩尔 3
Mouse 鼠标 6

MP3 player MP3 播放器 2, 12, 21, 43—44
Moore's law 摩尔定律 3, 12—13
Morris Worm 莫里斯蠕虫病毒 66
Musee d'Orsay 奥赛博物馆 96

N

Nanometre 纳米 29
NEC 日本电气公司 39
Netbook 上网本 2, 36
Netflix 美国奈飞公司 101
Netflix prize 奈飞奖 101
Neural network 神经网络 120—122
News blog 新闻博客 86
Nuclear experiment simulation 核试验模拟 57
NP-hard problem NP 难题 51—52, 123

O

O'Hara, Kieron 吉隆・奥哈拉 38
One-time pad 一次性密码本 74
Online advertising 在线广告 22
Open data 开放数据 101—102, 104
Open-source software 开源软件 81—82
Operating system 操作系统 35, 82
Open Wet Ware "开放湿件" 项目 113
Optical lithography 光刻法 37
Optimization 最优化 18
Oscar predictions 奥斯卡预测 116
Outlook mail program 微软邮件程序 63, 109
Outsourcing 业务外包 85
Oxford Literary Festival 牛津文学节 98

P

Pan-STARRS 泛星计划 45
Password 密码 74
Personal Digital Assistant (PDA) 掌上电脑 34
Personal firewall 个人防火墙 67
Pfleeger, Charles 查尔斯・弗莱格 75
Pfleeger, Shari 莎丽・弗莱格 75
Photo-mask 光掩模 29
Pinsent Masons 品诚梅森律师事务所 85
Plain text 纯文本 70—71
Pogue, David 戴维・波格 44
Post Office teletype 邮局电传打字机 2
Prediction market 预测市场 115
Private key 个人密钥 72—73
Processor 处理器 4, 50
Programmable Read-Only Memories (PROM) 可编程只读存储器 32
Programmableweb.com "可编程网络" 网站 105
Project Gutenberg 古腾堡计划 97
ProLoQuo2Go 一种为有语音差异的人设计的语音合成器应用程序 44
Public key 公开密钥 72—73
Public key cryptography 公开密钥加密 72

Q

Quantum computing 量子计算 22—23, 122—124

R

Rabbit 兔子电脑病毒 66
Radamacher, Paul 保罗·拉德马赫 106
Read Only Memory (ROM) 只读存储器 32
Read Write Memory (RWM) 读写存储器 32
Reflect tool 映射工具 48
Resist 抵制 31
Retina 一种帮助色盲人士辨别物体颜色的应用程序 44
RFID tag 射频识别标签 15, 21, 42—43
Rivest, Ronald 罗纳德·李维斯特 72
Royal Dirkzwager 荷兰皇家德克兹瓦格公司 39

S

San Diego Supercomputer Centre 圣迭戈超级计算机中心 46
Secure sockets layer 安全套接层 73—74
Set partition problem 集合划分问题 51
Shadbolt, Nigel 奈杰尔·沙德博尔特 38
Shamir, Adi 阿迪·萨莫尔 72
Simon, John 约翰·西蒙 91
Single-chip Cloud Computer 单芯片云计算机 120
SINTEF Group 辛泰夫集团 20
Social engineering 社会工程 76, 78
Sony Walkman 索尼随身听 12
Spamming 大量发送垃圾邮件 68
Spoofing attack 电子攻击 21
Static Random Access Memory (SRAM) 静态随机存取存储器 34
Statoil 国家石油公司 15
Stross, Randall 兰德尔·斯特罗斯 114
Subatomic erosion 亚原子侵蚀 36
Substrate 基底 30
Supercomputer 超级计算机 55—56
Swarm optimization 群优化 18—19, 53
Symmetric cipher 对称加密 71

T

Tanenbaum，Andrew 安德鲁·塔南鲍姆 82
Ted Conference 泰德会议 44
The Bookseller《书商》周刊 96
The long tail 长尾效应 93
Therac-25 25 号医用直线加速器 19
Time bomb 定时炸弹 66
Thomas Watson 托马斯·沃森 3
Torvalds, Linus 林纳斯·托瓦兹 82
Transistor 晶体管 28, 32
Trapdoor 陷阱门 66
Travelling salesman problem 旅行推销员问题 52, 125
Trojan horse 特洛伊木马病毒 14, 21, 65
Truth table 真值表 26
Tweet 推文 112
Twitter 推特 104, 112

U

Uhear 一个可以测试个人听力的应用

程序 44
USB memory USB 存储器 33
US Geological Survey 美国地质调查局 104

V

Verichip 无线射频识别芯片 43
Very Large Scale Integration (VLSI) 超大规模集成电路 20, 29, 52—53, 122
Video podcast 视频播客 88
Virgin Records 维京唱片公司 80
Virus 病毒 63—69
Virus signature 病毒签名 67
Virus checker 病毒检测程序 21
von Neumann architecture 冯·诺伊曼体系结构 117—118
von Neumann, John 约翰·冯·诺伊曼 117

W

Weather forecasting 天气预报 57
Web server 网络服务器 8—9, 36
Weinberg, David 戴维·韦恩伯格 121
Wicked problems 棘手问题 51, 53
Wikipedia 维基百科 82, 96, 102—104, 111, 113
WikiWikiWeb 维基维基网 110—111
Williamson, Malcolm 马尔科姆·威廉姆森 72
Wordia 一个在线视觉字典网站 17
World Wide Web 万维网 8
Wu, Tim 吴修铭 128

X

X-Box 微软的视频游戏机 127

Y

Yahoo 雅虎 107
Yahoo Pipes "雅虎管道" 聚合工具 107
YouTube "优兔" 视频网站 112

Z

Zavvi "扎维" 音像制品网站 80
Zittrain, Jonathan 乔纳森·齐特林 61, 126—127
Zombie Computer 僵尸电脑 14—15

Darrel Ince

THE COMPUTER

A Very Short Introduction

Contents

List of illustrations

Chapter 1
The naked computer

Introduction

One of the major characteristics of the computer is its ability to store data. It does this by representing a character or a number by a pattern of zeros and ones. Each collection of eight zeros and ones is known as a 'byte'; each individual one or zero is known as a 'bit' (binary digit). Computer scientists use various terms to describe the memory in a computer. The most common are the kilobyte, the megabyte, and the gigabyte. A kilobyte is 10^3 bytes, a megabyte is 10^6 bytes, and a gigabyte is 10^9 bytes.

The first computer I used was an Elliot 803. In 1969, I took a computer-programming course at my university which used this computer. It was situated in a room which was about 40 foot by 40 foot, with the hardware of the computer contained in a number of metal cabinets, each of which would fill almost all of the en-suite bathroom I have at home. You submitted your programs written neatly on special paper to two punch-tape operators, who then prepared a paper-tape version of the program. Each row of the paper tape contained a series of punched dots that represented the individual characters of the program.

The program was then taken to the computer room, the tape read by a special-purpose piece of hardware, and the results displayed on a device known as a Post Office Teletype; this was effectively a typewriter that could be controlled by the computer, and it produced results on paper that were barely of better quality than toilet paper.

The storage capacity of computers is measured in bytes; the Elliot computer had 128 thousand bytes of storage. It used two cabinets for its memory, with data being held on small metallic rings. Data were fed to the computer using paper tape, and results were obtained either via paper or via a punch which produced paper tape. It required an operator to look after it, and featured a loudspeaker which the operator could adjust in volume to check whether the computer was working properly. It had no connection to the outside world (the Internet had yet to be invented), and there was no hard disk for large-scale storage. The original price of the first wave of Elliot 803s was £29,000, equivalent to over a hundred thousand pounds today.

While I am writing this chapter, I am listening to some Mozart on a portable music device known as an MP3 player. It cost me around £180. It comfortably fits in my shirt pocket and has 16 gigabytes of memory – a huge increase over the capacity of the only computer at my old university.

I typed this book on a computer that was known as a netbook. This was a cut-down version of a laptop computer that was configured for word processing, spreadsheet work, developing slide-based presentations, and surfing the Internet. It was about 10 inches by 6 inches. It also had 16 gigabytes of file-based memory used for storing items such as word-processed documents, a connection to the Internet which downloaded web pages almost instantaneously, and a gigabyte of memory that was used to store temporary data.

There is clearly a massive difference between the Elliot 803 and the computers I use today: the amount of temporary memory, the amount of file-based memory, the processing speed, the physical size, the communications facilities, and the price. This increase is a testament to the skills and ingenuity of the hardware engineers who have developed silicon-based circuits that have become smaller and more powerful each year.

This growth in power of modern computers is embodied in a law known as 'Moore's law'. This was expounded by Gordon Moore, the founder of the hardware company Intel, in 1965. It states that the density of silicon circuits used to implement a computer's hardware (and hence the power of a computer) will double every eighteen months. Up until the time of writing, this 'law' has held.

The computer has evolved from the physical behemoths of the 1950s and 1960s to a technological entity that can be stored in your jacket pocket; it has evolved from an electronic device that was originally envisaged as something only large companies would use in order to help them with their payroll and stock control, to the point where it has become an item of consumer electronics as well as a vital technological tool in commercial and industrial computing. The average house will contain as many as 30 computers, not only carrying out activities such as constructing word-processed documents and spreadsheet tables, but also operating ovens, controlling media devices such as televisions, and regulating the temperature of the rooms.

Even after 70 years, the computer still surprises us. It surprised a number of computer scientists in the fifties, who predicted that the world only needed a small number of computers. It has surprised me: about 20 years ago, I saw the computer as a convenient way of reading research documents and sending email, not as something that, combined with the Internet, has created a global community that communicates using video technology, shares photographs, shares video clips, comments on news, and reviews books and films.

Computer hardware

One aim of this book is to describe how the computer has affected the world we live in. In order to do this, I will describe the technologies involved and the applications that have emerged over the last ten years – concentrating on the applications.

First, the basic architecture of the computer; I will describe this architecture in a little more detail in Chapter 2. This is shown in Figure 1. The schematic shown in this figure describes both the earliest computers and the newest: the basic architecture of the computer has not changed at all over 60 years.

At the heart of every computer is one or more hardware units known as processors. A processor controls what the computer does. For example, it will process what you type in on your computer's keyboard, display results on its screen, fetch web pages from the Internet, and carry out calculations such as adding two numbers together. It does this by 'executing' a computer program that details what the computer should do, for example reading a word-processed document, changing some text, and storing it into a file.

Also shown in Figure 1 is storage. Data and programs are stored in two storage areas. The first is known as main memory and has the property that whatever is stored there can be retrieved very quickly. Main memory is used for transient data – for example, the result of a calculation which is an intermediate result in a much bigger calculation – and is also used to store computer programs while they are being executed. Data in main memory is transient – it will disappear when the computer is switched off.

Hard disk memory, also known as file storage or backing storage, contains data that are required over a period of time. Typical entities that are stored in this memory include files of numerical data, word-processed documents, and spreadsheet tables. Computer programs are also stored here while they are not being executed.

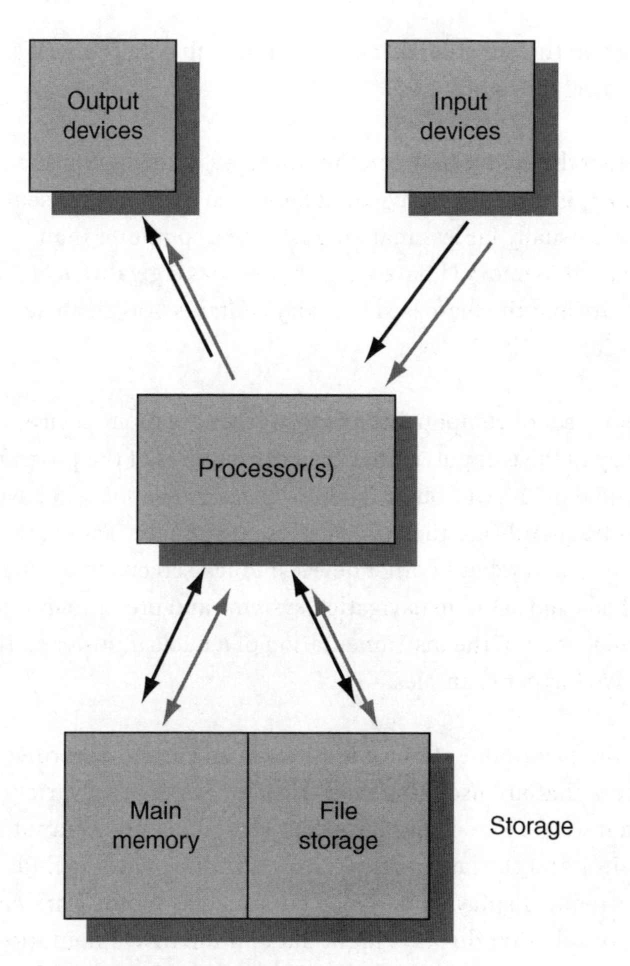

1. The architecture of a computer

There are a number of differences between main memory and hard disk memory. The first is the retrieval time. With main memory, an item of data can be retrieved by the processor in fractions of microseconds. With file-based memory, the retrieval time is much greater: of the order of milliseconds. The reason for this is that main memory is silicon-based and all that is required to read data there is for it to be sent along an electronic circuit. As you will see later, hard disk memory is usually mechanical and is stored on the metallic surface of a disk, with a mechanical arm retrieving the data.

Another difference between the two types of memory is that main memory is more expensive than file-based memory; consequently, there is usually far less main memory in a computer than file-based memory (I have a laptop that has 3 gigabytes of main memory and the file-based memory contains 500 gigabytes of storage).

Another set of components of a computer are input devices. These convey to the computer what the user requires of the programs executed by the computer. The two devices that you will have met most frequently are the keyboard and the mouse. There are, however, a number of other devices: touch screens that you find on iPods and satellite navigation systems and pressure monitors found as part of the instrumentation of a nuclear power station are two further examples.

The final component of a computer is one or more hardware devices that are used to display results. There are a variety of such devices. The most familiar to you will be the computer monitor and the laser printer; however, it can also include advertising displays found at events such as football matches, the console that displays flight data on the instrumentation found in the cockpit of a plane, the mini-printer that is used to produce a supermarket receipt, and the screen of a satellite navigation device.

The working definition of a computer that I shall use within this book is:

> A computer contains one or more processors which operate on data. The processor(s) are connected to data storage. The intentions of a human operator are conveyed to the computer via a number of

input devices. The result of any computation carried out by the processor(s) will be shown on a number of display devices.

You may think this statement is both pedantic and self-evident; however, I hope that you may see as this book unfolds that it has a number of radical interpretations.

Before leaving this section, it is worth looking at another indicator of the growth in power of computers. In their excellent book *The Spy in the Coffee Machine*, O'Hara and Shadbolt describe the progress made in computer-based chess. To be good at chess requires you to look ahead a number of moves and evaluate what your opponent would do for each of these moves, and then determine what move you would make for each of these moves, and so on. Good chess players hold lots of data in their heads and are able to carry out fast evaluations. Because of this, the computer has always been seen as potentially a good chess player.

The chess programs that have been written effectively store lots of moves and countermoves and evaluate them very quickly. O'Hara and Shadbolt describe how in 1951 a computer could only think ahead two moves, in 1956 a computer could play a very restricted game of chess on a smaller board but would take upward of 12 minutes to make a move. However, in 1997 a computer beat the world champion Gary Kasparov. This progress is partly due to improvements in software techniques for game playing; the major reason though is that computers have become faster and faster.

The Internet

Computers do not operate in isolation: most are connected to a computer network. For most computers, this will be the huge collection of computers and communication facilities known as the Internet; however, it could be a network that controls or monitors some process, for example a network of computers that keep a plane flying, or a network of computers used to monitor the traffic flow into and out of a city.

The Internet has had a major effect on the way computers are currently being used; so it will be worthwhile looking briefly at how it interacts with a typical computer – say the PC that you use at home.

The Internet is a network of computers – strictly, it is a network that joins up a number of networks. It carries out a number of functions. First, it transfers data from one computer to another computer; to do this, it decides on the route that the data takes: there is a myth that when you carry out some activity using the Internet, for example downloading a web page, the connection between the computer holding the page and your computer is direct. What actually happens is that the Internet figures out a route that the data takes via a number of intermediate computers and then routes it through them. So when you see a web page displayed on your computer, that page may have been split into blocks of data, with each block having travelled through a number of continents and traversed a number of intermediate computers belonging to companies, universities, charitable organizations, and government organizations.

The second function of the Internet is to enforce reliability. That is, to ensure that when errors occur then some form of recovery process happens; for example, if an intermediate computer fails then the software of the Internet will discover this and resend any malfunctioning data via other computers.

A major component of the Internet is the World Wide Web; indeed, the term 'Internet' is often used synonymously with the term 'World Wide Web'. The web – as I shall refer to it from now on – uses the data-transmission facilities of the Internet in a specific way: to store and distribute web pages. The web consists of a number of computers known as *web servers* and a very large number of computers known as *clients* (your home PC is a client). Web servers are usually computers that are more powerful than the PCs that are normally found in homes or those used as office computers. They will be maintained by some enterprise and will contain individual web pages relevant to that enterprise; for example, an online book store such as Amazon will maintain web pages for each item it sells.

The program that allows users to access the web is known as a *browser*. When you double-click the browser icon on your desktop, it will send a message to the web asking for your home page: this is the first page that you will see. A part of the Internet known as the Domain Name System (usually referred to as DNS) will figure out where the page is held and route the request to the web server holding the page. The web server will then send the page back to your browser which will then display it on your computer.

Whenever you want another page you would normally click on a link displayed on that page and the process is repeated. Conceptually, what happens is simple. However, it hides a huge amount of detail involving the web discovering where pages are stored, the pages being located, their being sent, the browser reading the pages and interpreting how they should be displayed, and eventually the browser displaying the pages.

I have hidden some detail in my description. For example, I have not described how other web resources such as video clips and sound files are processed. In a later chapter, I will provide a little more detail. At this point, it is just worth saying that the way that these resources are transferred over the web is not that different to the way that web pages are transferred.

The Internet is one of the major reasons why computers have been transformed from data-processing machines to a universal machine that can, for example, edit music files, predict the weather, monitor the vital signs of a patient, and display stunning works of art. However, without one particular hardware advance the Internet would be a shadow of itself: this is broadband. This technology has provided communication speeds that we could not have dreamed of 15 years ago. Most users of the Internet had to rely on what was known as a dial-up facility which transferred data at around 56 kilobits of data a second. When you consider that the average web page size is around 400 kilobits, this means it would take around 7 seconds for a web page to be displayed in your browser. In the 1990s, companies used dedicated communications hardware to overcome this lack of speed.

Unfortunately, the average user was unable to do this until broadband became generally available.

Typical broadband speeds range from one megabit per second to 24 megabits per second, the lower rate being about 20 times faster than dial-up rates. As you will see later in the book, this has transformed the role of the home-based computer.

Software and programs

The glue that binds all the hardware elements shown in Figure 1 together is the computer program. When you use a word processor, for example, you are executing a computer program that senses what you type in, displays it on some screen, and stores it in file-based memory when you quit the word processor. So, what is a computer program?

A computer program is rather like a recipe. If you look at a recipe in a cookbook, you will see a list of ingredients and a series of instructions that will ask you to add an ingredient, mix a set of ingredients together, and place a collection of ingredients in an oven. A computer program is very much like this: it instructs the computer to move data, carry out arithmetic operations such as adding a collection of numbers together and transfer data from one computer to another (usually using the Internet). There are, however, two very important differences between recipes and computer programs.

The first difference is size. While a typical recipe might contain about 20 lines of text, computer programs will contains hundreds, thousands, or even millions of lines of instructions. The other difference is that even a small error can lead to the catastrophic failure of a program. In a recipe, adding four eggs instead of three may result in a meal with a slightly odd taste or texture; however, mistyping the digit 1 instead of 2 in a million-line program may well result in a major error – even preventing the program running.

There are a variety of programming languages. They are categorized into high-level and low-level languages. A high-level language, such as Java or C#, has instructions that are translated to many hundreds of the individual instructions of the computer. A low-level language often has a one-to-one relationship with the basic computer instructions and is normally used to implement highly efficient programs that need to respond to events such as the temperature of a chemical reactor becoming critical.

Every few days the media features a story about a software project that has overrun or is over budget or a computer system that has dramatically failed. Very rarely can these failures be attributed to a failure of hardware. The failures occur for two reasons. The first reason for a malfunctioning of an existing computer system is technical error, for example a programming error that was not detected by testing. The second reason is due to managerial failings: projects that overrun or dramatically exceed their budgets tend to occur because of human factors, for example a poor estimate of project resources being produced or a customer changing their mind about the functions a system should implement.

My own view is that given the complexity of modern computer systems, it is hardly surprising that projects will be late and that errors will be committed by developers.

Book themes

The first theme of the book is how hardware advances have enabled the computer to be deployed in areas which would have been unheard of a decade ago. The circuit that a computer processor is deposited on can be easily held in the palm of one hand rather than in a large metal cupboard. A memory stick containing 16 gigabytes of data can easily be attached to a key-ring. Moore's law implies that the computational power of a computer processor doubles every two years. You can now buy hard disk storage of 500 gigabytes for under £60. There are a number of implications. The first is that in the past decade computers have been able to do things few people dreamt of in the 1990s, for example British Telecom's Vision programme that brings television over the Internet. The second is that the reduction in size of computer hardware has enabled them to be physically deployed in environments which would have been impossible a few years ago.

The second theme is how *software* developers have taken advantage of advances in hardware to produce novel applications. An example of this is that of MP3 players such as the Apple iPod. The iPod, and other devices such as the Sony Walkman, obviously rely on advances in hardware. However, they also rely on a software-based technique which, when applied to a sound file, compresses it so that it occupies 10% of its original size without an appreciable decline in sound quality.

A third theme is how the Internet has enabled computers to be connected together in such a way that they behave as if they were just one big computer. This is embodied in an idea known as 'cloud computing' whereby data, rather than being stored in a local database, are held in a number of computers connected to the Internet and may be accessed by programs that can be developed by Internet users who have relatively low-level programming skills.

Allied to this idea is that of the Internet as a huge resource of data which the computer user can tap into. This includes data such as that published by the British government's data.gov.uk and US government's Data.gov programs, but also data that have been contributed directly or indirectly by users of the Internet. For example, there are sites that enable you to home in to your town or village and examine the broadband speeds that are being experienced by your neighbours, the data that these sites contain having been donated by the users of the site.

A fourth theme is how the Internet has provided creative facilities that were only available to professionals. For example, computer hardware advances, software advances, and advances in the technologies used to create video cameras mean that anyone can become a film director and display their results on the Internet. A computer user can now buy hardware and software for less than a thousand dollars that enables them to reproduce the features of a recording studio of the 1990s.

A fifth theme is how advances in computer processor hardware have enabled number-crunching applications which, until a few years ago, were regarded as outside the realm of computation. Moore's law implies that computer processors become twice as powerful as they were eighteen months previously. The consequence of this is that over the past decade, processors have become many times more powerful and, combined with other hardware improvements such as the increased speed of data-storage devices means that, for example, simulations involving the natural world – for example, simulations of hurricanes – can now be easily carried out without deploying powerful supercomputers.

A sixth theme is how the computer has become a disruptive technology in that it has both transformed and eliminated many skills. An example here is photography. When I visit a tourist site, I hardly ever see film cameras being used: almost invariably, small palm-sized digital cameras are now the norm. Moreover, the photographs that are taken can be brought home in a memory stick, placed in a home computer, and then printed. Relatively cheap programs such as Adobe Photoshop can now be used to improve these photographs by, for example, adjusting the exposure.

No longer does the development of a photograph involve the dousing of a film in chemical baths in a darkroom. This is clearly an improvement; however, there is another side to the coin which has resulted in fewer job opportunities for photographers. There is a web site known as Flickr. This is a photo-sharing site where Internet users upload photographs and display them to visitors to the site. Newspaper editors who want cheap stock photographs for an issue of their paper (for example, a picture of a robin for a Christmas edition) can purchase such a picture for a fraction of the amount that they would have to pay a freelance photographer.

A seventh theme is that of the insecure computer. A computer that stands alone with no connections to a network is perfectly safe from any technological attack; the only threat that the owner of the computer should be aware of is that of having it stolen. However, very few computers are in this state: most are connected to the Internet. This means that they become prone to a large variety of attacks, from those that create a mild nuisance effect to serious attacks which can completely stop a computer from working. An example of this is the zombie computer. This is a computer attached to the Internet that has been compromised by a hacker, a computer virus, or a Trojan horse.

The most common use for a zombie computer is to act as a mail server and send spam email; this is email that tries to sell you something you don't need (Viagra, cheap stocks and shares, or pornographic publications, for example) or attempts to steal information such as your bank account identity. Most owners of such computers are unaware that their system is being used in this way. It is because the owner tends to be unaware that they are referred to as 'zombies'. In May 2009, the security company McAfee estimated that there were around 12 million new zombies attached to the Internet. This is really quite an extraordinary figure for a computer infestation.

Some examples

Before looking at these themes in more depth, it is worth examining some examples of the themes in action.

The Norwegian oil company StatOil uses blue mussels to monitor any leaks around their oil rigs. When there is an oil leak, the mussels contract their shells. Concerned with the environmental and revenue impacts of leaks during oil drilling, StatOil sought a way to replace a manual process that involved submersible vehicles and included deep-sea divers. What they did was to attach RFID tags to the shells of blue mussels. These are small, silicon-based, data-holding chips that also include a computer. When the blue mussels sense an oil leak, they close; this makes the RFID tags emit signals that indicate this event has occurred; these signals are picked up by a computer on the rig which then shuts down the activity that was causing the leak. For example, if drilling were taking place, the drilling line would be automatically stopped. This unusual application is possible as a consequence of advances in the miniaturization of computer circuits.

Google Inc. is the company that hosts the hugely popular search engine known as Google. One of the things that the search engine does is to store the queries made by the users so, for example, you can visit a Google web site and discover which are the most popular queries. In 2008, the fastest-rising queries from the

United Kingdom were: 'iPlayer', 'facebook', 'iphone', 'youtube', 'yahoo mail', 'large hadron collider', 'Obama', and 'friv'. Most of these terms are associated with hugely popular web sites or electronic devices such as the iPhone. The last entry, 'friv', is an online games site.

As you will see later, a huge amount of information can be obtained from the queries that users submit to a search engine. It is now common practice for police investigators to explore the use of a search engine by a suspected murderer. In a murder where the victim's neck was broken, they would check for search terms such as 'neck', 'snap', 'break', 'rigor mortis', and 'body decomposition' which the murderer might have submitted to the search engine.

An interesting application of the massive amount of stored data that Google keeps of the queries that are made is in tracking influenza. Two Google engineers tracked the incidence of queries such as 'thermometer', 'flu symptoms', 'muscle aches', and 'chest congestion', and compared the location of the Internet users who made the queries against the US Center for Disease Control database and discovered a very close correlation: in effect, they discovered that the volume of queries involving the search words was similar to the density of flu cases. You can now access a web site managed by Google Inc. that shows the growth of flu cases in a number of countries over a period of time.

This is an example of a major theme of this book: that of the computer not only having access to the data on its own hard drive, but also to the massive amount of data stored on the computers spread around the Internet.

Another example of a use of computers beyond the limited visions of the 1970s and 1980s concerns the way that computers are connected together in order to work collaboratively.

Researchers in the applied sciences have for the past 20 years tried to squeeze processing power from their computers. For example, the Human Genome Project has mapped the gene structure of humankind and researchers are now using this information to detect the genetic causes for many types of illness. This work has required the use of expensive supercomputers that contain a large number of processors. However, a number of researchers in this area, and in other areas such as climatology, have come up with the novel idea of asking the public to run processor-intensive programs.

A good example of this is Folding@home. This project looks at the structure of proteins in order to detect therapeutic regimes that can be used for treating patients with conditions such as Alzheimer's disease. Researchers involved in this project have enlisted around 30,000 home computers to spread the computational load. Volunteers use their spare processor and memory capacity to take a small part of a computer program that carries out protein simulation and produce results that are fed back to a master computer that coordinates the processing.

This is not the only application of a technique known as 'mass computing' or 'mass collaboration'. There are projects that attempt to analyse the radio waves from outer space in order to discover whether there is intelligent life beyond our universe, those that simulate atomic and sub-atomic processes, and many projects associated with molecular biology. In the past, supercomputers containing a large number of processors situated in a small number of research institutes were used – and indeed are still used – however, hardware advances and the increasing availability of the broadband Internet has meant that we can all participate in major research projects with little effect on our home computers.

wordia is a visual dictionary that anyone can access via their home computer. It contains words of course, but each word is accompanied by a video of someone telling you what the word means to them. It is one of the most delightful web sites that I have come across and is an example of the phenomenon known as mass collaboration in action. This is an example of an application related to a major theme of the book: that of the computer being part of a loosely coupled global computer.

Another example involves the construction of computer circuits. As engineers try to squeeze more and more electronic components onto silicon chips, the design of such chips becomes much more difficult: for example, placing two metallic connections close to each other would cause electrical interference that would result in the circuit malfunctioning. Given that millions of such circuits might be manufactured and embedded in computers, a mistake would be hugely expensive for the manufacturer. The complexity of design is such that the only viable way to develop the architecture of a silicon-based circuit is by using the computer itself.

The programs that are used to design computer circuits try to optimize some design parameter; for example, one class of programs attempts to squeeze connections on a silicon chip in such a way that the maximum number of connections are deposited subject to a number of constraints: that connections are not too close to each other and that the heat dissipation of the circuit does not rise past some threshold which would affect the reliability of the circuit. There are a number of techniques used for optimization; one recent very efficient class of programs is based on animal and insect behaviour.

An example of this is a technique known as 'swarm optimization' in which a number of computer processes collaborate with each other in order to discover an optimal solution to a problem using the simple mathematics used to describe how shoals of fish or flocks of birds behave. Here is an example of another theme of this book: the ingenuity of the programmer combined with hugely increased speeds that enable complex tasks to be carried out that even a small number of years ago would have been impossible to even think about.

Swarm optimization is an example of the revolution in the use of computers that has happened over the past two decades: it is represented by the progression of the computer that just carries out mundane processing steps such as calculating a wage bill to applications such as designing computers and controlling the inherently unstable fighter planes that have become the norm in our armed services.

So far, I have concentrated on computers and their uses that are visible. There are many, many more applications where the computer is not seen. My nearest city is Milton Keynes. When I drive to the city and then around its severely practical road system, I pass many, many unseen applications of the computer. I pass a speed camera controlled by a small microprocessor; a company that fabricates electronic equipment using robots controlled by a computer; the street lighting controlled by a very small, primitive computer; the Milton Keynes Hospital where most of the monitoring equipment that is used could not function without an embedded computer; and the shopping centre, where computers are used to keep the environment within each shop strictly controlled.

Increasingly, computers are being used in hidden applications where failure – either hardware or software failure – could be catastrophic and, indeed, has been catastrophic. For example, the Therac-25 was a computer-based radiation therapy machine which had a number of software problems. In the late 1980s, a number of patients received massive overdoses of radiation because of problems with the computer interface.

An example of a hidden application where failure can be catastrophic and which is an example of another theme is that of the control of an oil rig. A functioning oil rig draws extremely flammable oil or gas out of the earth, burns some of it off and extracts unusable by-products such as hydrogen sulphide gas from the oil. In ocean-based installations, this required a large number of human operators. Increasingly, however, computers are being used to carry out tasks such as controlling the flow of oil or gas, checking that there is no spillage, and regulating the burn-off process.

There have been instances of IT staff hacking into the software systems that are, for example, used to monitor oil rig operations, either for financial gain or because they had become disgruntled. What is not realized is that although computer crime such as spreading viruses is still prevalent there are whole collections of applications of computers that are just as vulnerable to attack. The SINTEF Group, a Norwegian think tank, has reported that offshore oil rigs are highly vulnerable to hacking as they move to less labour-intensive, computer-controlled operations – for example, the wireless links that are used to remotely monitor the operation of a rig and to keep a rig in its position via satellite navigation technology are particularly vulnerable.

This book

Each of the chapters of this book is centred about a theme that I have outlined in this chapter.

'The Small Computer' will describe how a computer architecture is mapped to silicon and the problems that the computer designer has to face when pushing more and more electronic components onto a piece of silicon. Topics that will be discussed here and in other chapters include: very large-scale integration, silicon fabrication, the hardware design process, and new techniques and technologies for design such as the use of artificial intelligence programs to maximize or minimize some factor such as heat dissipation.

'The Ubiquitous Computer' will describe how miniaturization has led to the computer becoming embedded into a variety of electronic and mechanical devices. Examples discussed in this chapter and other chapters: RFID tags, the use of supermarket loyalty cards, computers used for the monitoring of the infirm or elderly, wearable computers, computers used within virtual reality systems, and the convergence that is occurring between the phone, the MP3 player (iPod), and the computer.

'The Global Computer' will look at how the Internet has enabled large numbers of computers to be connected together in such a way that they can be used to attack wicked problems – that is, problems that are computationally very difficult to solve. The chapter starts by looking at one particular computationally wicked application from genome sequencing. I will then describe a concept known as grid computing where very large numbers of computers are connected together in such a way that their spare capacity can be used to attack such hard problems. The chapter will conclude by looking ahead to the seventh chapter and briefly describe how the grid concept has become commercialized into something known as cloud computing. This involves regarding the Internet as just one huge computer with almost infinite computing power and data-storage facilities.

'The Insecure Computer' looks at some of the threats – both technological and human – that can result in major security problems. The chapter will cover the whole panorama of attacks including virus attacks, Trojan attacks, denial of service attacks, spoofing attacks, and those that are caused by human error. The chapter will look at the defences that can be employed, including firewalls, intrusion detectors, virus checkers, and the use of security standards. A strong point that I make is that technological defence is not enough but that it has to be melded with conventional security controls.

'The Disruptive Computer' describes how the computer has had a major disruptive effect. Most of the examples describe disruption engendered by the combination of the computer and the communications technologies employed in the Internet. It will examine how in media industries, for example newspapers, have declined over the last five years and how online advertising has eaten into the revenues of television companies. The concluding part of the chapter will examine a number of areas where computers have de-skilled, transformed, moved, or eliminated certain jobs.

'The Cloud Computer' describes how the Internet has enabled not just developers but moderately skilled individuals to treat this network like a massive computer. A number of companies such as Amazon provide public access to huge product databases and programming facilities in such a way that applications can be developed that mesh across a number of areas. This has led to the concept of the cloud computer: huge numbers of processors and databases connected by the Internet with software interfaces that anyone can use. The chapter introduces the idea of software mashing: the process whereby sophisticated applications can be constructed by integrating or 'mashing' large chunks of existing software.

'The Next Computer' is a relatively short chapter. It looks at some of the blue-skies work that is being carried out by researchers in an attempt to overcome the limitations of silicon. It will focus on quantum computing and biological computing. A quantum computer is a computer that carries out its processes using quantum effects such as entanglement to operate on data. It is very early days as yet, but theoretical studies and some early experiments have indicated that huge processing speed-ups are possible with quantum computers.

The effect of such computers could be devastating. For example, much of commercial computing depends on cryptographic techniques that rely on the huge computational complexity of certain classic number processing algorithms. Quantum computers may be capable of making these algorithms solvable and hence open up the Internet to easy attack.

The chapter will also describe the principles behind the DNA computer. This is a half-way house between current computer technology and quantum computers. DNA computers use the genetic properties of biological strands to provide very large parallel processing facilities. Effectively, DNA computers implement a large number of hardware processors which cooperate with each other to solve hard computational problems.

A major idea I hope to convey to you in Chapters 4 and 7 is that regarding the computer as just the box that sits on your desk, or as a chunk of silicon that is embedded within some device such as a microwave, is only a partial view. The Internet – or rather broadband access to the Internet – has created a gigantic computer that has unlimited access to both computer power and storage to the point where even applications that we all thought would never migrate from the personal computer are doing just that.

An example of this is the migration of office functions such as word processing and spreadsheet processing – the bread and butter of many home computers. Google Inc. has launched a set of office tools known as Google Apps. These tools are similar to those found in Microsoft Office: a word processor, a spreadsheet processor, and presentation package similar to PowerPoint etc.

Chapter 2
The small computer

Introduction

The last 30 years has seen an amazing improvement in the capability of computers in terms of their processing speed, size of memory, cost, and physical size. Processors have increased their power from around 90 kIPS in the early 1970s to many thousands of MIPS in the second decade of the 21st century. The speed of a processor is expressed in instructions per second (IPS) where an instruction is some action that the computer takes, for example adding two numbers together; the prefix 'k' stands for a thousand, while the prefix 'M' stands for a million.

Memory capacity has also increased: the Elliot 803 computer that I described in the previous chapter contained 128 k bytes of memory held in a cabinet the size of a dozen coffins: my iPod contains 16 gigabytes of storage.

How has this increase in speed and memory capacity happened? In this chapter, I answer this question; however, before doing so, it is worth looking briefly at how data and computer programs are held in the computer.

The binary number system

We are all used to the decimal system of numbering. When you see a number such as 69126, what it stands for is the number

$$6\times10^4+9\times10^3+1\times10^2+2\times10^1+6\times10^0$$

where each digit represents the result of multiplying itself by a power of ten (any number raised to the power 1, for example 10^1, is always itself, in this case 10, and any number raised to the power zero is always 1).

We say that the base of a decimal number is 10; this means that we can express any decimal number using a digit between 0 and 9. With binary numbers, the base is 2; this means that we can interpret a binary number such as 11011 as

$$1\times2^4+1\times2^3+0\times2^2+1\times2^1+1\times2^0$$

and it will have the decimal value 27 (16+8+0+2+1).

Numbers are stored in binary form in this way. Text is also stored in this form as each character of the text has an internal numeric equivalent. For example, the American Standard Code for Information Interchange (ASCII) is a standard used throughout computing to designate characters. The code assigns a numeric value for each character that can be stored or processed by a computer – for example, the capital A character is represented by a binary pattern equivalent to the decimal number 65.

The binary system is also used to represent programs.
For example, the pattern

1001001110110110

might represent an instruction to add two numbers together and place them in some memory location.

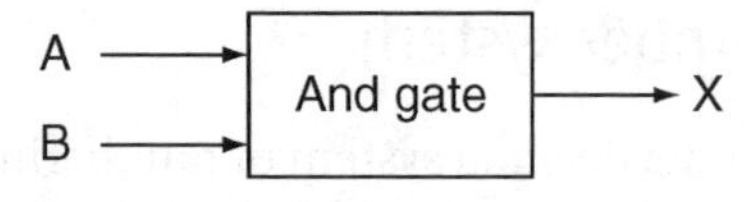

2. A schematic of an And gate

Computer hardware

A computer will consist of a number of electronic circuits. The most important is the processor: this carries out the instructions that are contained in a computer program. As you will remember from the preceding chapter, there are two types of memory: main memory used to store relatively small quantities of data and file-based memory which is used to store huge amounts of data such as word-processor files.

There will also be a number of other electronic circuits in a computer: for example, if you look at the back of a desktop computer, you will often see a black rod attached to a metal strip. The rod contains the aerial that picks up wireless signals from a device known as a modem that connects to the Internet. Behind the strip will be a circuit that converts the signals that are picked up by the aerial into a form that can be used by the computer so that, for example, it can display a web page on your monitor.

There are a number of individual circuit elements that make up the computer. Thousands of these elements are combined together to construct the computer processor and other circuits. One basic element is known as an And gate, shown as Figure 2. This is an electrical circuit that has two binary inputs A and B and a single binary output X. The output will be one if both the inputs are one and zero otherwise. This is shown in a tabular form known as a truth table; the truth table for the And gate shown in Figure 2 is shown in Table 1.

Table 1. The actions of an And gate

A	B	X
0	0	0
0	1	0
1	0	0
1	1	1

There are a number of different circuits inside the computer – the And gate is only one example – when some action is required, for example adding two numbers together, they interact with each other to carry out that action. In the case of addition, the two binary numbers are processed bit by bit to carry out the addition.

So, how does a computer do its work? The best way to describe this is to outline what happens when I use a word processor. When I click the MS Word icon on my desktop, the Windows operating system senses the click and then loads the MS Word word processor into the main memory of the computer.

The program then starts executing. Each time I carry out some action the word-processor program senses it and part of its program code is executed. The execution is carried out in what is known as the fetch-execute cycle. Here the processor fetches each programming instruction and does what the instruction tells it to do. For example, an instruction may tell the computer to store what I have typed in a file, it may insert some text into some part of the word processed document, or it may quit the word processor.

Whatever actions are taken by a program such as a word processor, the cycle is the same; an instruction is read into the processor, the processor decodes the instruction, acts on it, and then brings in the next instruction.

So, at the heart of a computer is a series of circuits and storage elements that fetch and execute instructions and store data and programs. Over the last 70 years, a variety of technologies have been used for constructing a computer. The very first computers were based on electrical relays; these were mechanical switches which would have two states: binary one would be represented by the relay being closed, while zero would be represented by a relay being open. When you hear a computer programmer talking about a 'bug' in their program, the term comes from the use of relay computers. In 1946, Grace Hopper, one of the pioneers of computer programming, joined the Computation Laboratory at Harvard University where she worked on early relay computers. She described how they traced a program error to a moth trapped in a relay, coining the term 'bug' for a software error.

The first real generation of computers used electronic circuits based around devices known as valves. These looked a bit like small light bulbs and could be switched electronically from a zero (off) to a one (on) state via a signal to the valve. Programmers communicated with these computers using paper tape or punched cards which either held data to be processed or the programs that carried out the processing.

The main memory that was used in early computers employed circular pieces of magnetic material known as cores. These stored binary one or binary zero depending on their magnetized state.

First-generation computers were succeeded by second-generation computers that used transistors. A transistor was a lump of silicon that could be switched on and off; the Elliot computer I described in the previous chapter relied on such a technology.

During the late 1960s and early 1970s, physicists, materials scientists, and electronic engineers managed to deposit the circuits that were implemented by transistors in second-generation computers onto silicon chips. This is a process known as Very Large Scale Integration (VLSI). These third-generation computers are the ones we use today. It is VLSI that has been the technology that has provided the incredible miniaturization, speed, and capacity of today's computers. Miniaturization, exemplified by the width between components, has decreased from around 1.00 μm in the early 1990s to 40 nm in the early part of the 21st century. The symbol μm stands for a millionth of a metre, and the symbol nm stands for a nanometre – one thousandth of one millionth of a metre.

Computer circuits

Modern computer hardware relies on silicon. There are a number of manufacturing steps that are carried out to transform a block of silicon into a processor or an interface circuit such as the one used to drive the monitor of a computer.

The first step is the growing of a single crystal of silicon as a cylinder. When the growth of the cylinder has been completed, circular slices known as 'wafers' are cut from the cylinder in the same way that you would cut slices from a cylinder of luncheon meat, the only difference being that the slices are usually a fair bit thinner than the meat. After the slices have been cut, they are polished.

The next step is to embed a design on each silicon wafer for the circuit that is to be implemented. This is done via a device known as a photo-mask. This is a grid that lays out the pattern of the circuit on the wafer together with the components of the circuit. Ultraviolet light is shone through the grid onto a portion of the wafer and this forms the guidelines for the circuit to be deposited on it. Normally a number of similar circuit plans are etched onto the silicon wafer.

In detail, the fabrication process proceeds as follows. First, the silicon wafer is baked in an oven filled with oxygen. This forms a thin layer of silicon dioxide on the surface. The wafer is then coated with another thin layer of an organic material known as a resist. So we now have a silicon base, often known as a substrate, a silicon dioxide layer, and a resist layer at the top.

Ultraviolet light is then shone through the mask onto a portion of the wafer. The structure of the resist is broken up by the light but the remaining layers are unaffected. The process is very similar to the way that a black and white photograph is developed. Once a portion of the wafer has had the light shone on it the mask is moved and the next part of the wafer has the pattern etched.

The next stage is for the wafer to be fabricated. This involves placing it along with many other wafers in a bath of solvent which dissolves those parts of the resist that have received the ultraviolet light.

The silicon wafer will now have a layer of silicon, a layer of silicon dioxide, and the part of the resist layer that has been unaffected by the ultraviolet light. The part of the wafer that has been removed by the solvent will have exposed areas of silicon dioxide. These are then removed by applying another solvent which will expose the underlying silicon.

The silicon wafer will now contain a layer of silicon parts which are exposed, a silicon dioxide layer which will have sections cut out of it exposing the silicon, and the resist which will have the same sections cut as the silicon dioxide.

The next step is to remove the resist by using a solvent that dissolves it. The wafer now contains a base layer of silicon with the circuit pattern etched in it. The exposed part of the silicon layer is then treated in order to make it capable of conducting electrical signals. The wafer now has treated silicon which represents the circuit and new silicon dioxide that acts as an insulator which ensures that signals that pass though one part of the silicon do not affect other signal paths.

Further layers are then deposited to complete the circuit with the last layer being one of silicon dioxide. This is etched with holes that enable connections to be made with the underlying circuits.

The process of packaging the circuits now starts. There are a variety of packaging techniques. I shall describe the simplest. First, square metallic connections known as pads are deposited around the edge of each circuit. Another layer of silicon dioxide is then placed on the wafer with holes etched in the layer to enable connections to be made to the pads.

Each circuit is then tested by a special purpose piece of electronic equipment which will engage with the pads and send signals to some of the pads and monitor the effect of the signals on other pads. Any circuit that fails the test is marked with a dye and eventually rejected. If the circuits pass the test, another layer of silicon dioxide or silicon nitride is placed over the circuit and connection holes made in the layer to the pads. This final layer acts as physical protection.

The final step is to cut each circuit from the silicon wafer. This is achieved by a mechanical cutter; this is similar in concept to the way that a glazier will cut a shape out of glass. The wafer has now become a collection of identical chips.

The final step is for each chip to be mounted in some sort of frame in order that it can be fitted into a computer. There are a variety of techniques for this. A simple one involves attaching the chip on a lead frame using an adhesive that helps conduct heat away from the chip and then placing signal wires on the chip that connect with the pads. Once the wires are added, the chip is covered in some plastic-based material as a final protection.

If you are interested in more details, the excellent book *BeBOP to the Boolean Boogie: An Unconventional Guide to Electronics* by Clive Maxfield is a great introduction to computer electronics (see Further reading).

Computer memory

There are two sorts of memory devices: read-only memory (ROM) devices and read-write memory (RWM) devices. The former hold data that cannot be altered; the latter can be erased and data rewritten.

Computer memory is implemented as silicon and is fabricated in the same way that hardware processors and other circuits are fabricated. The only difference between a computer memory and, say, the circuit that communicates with the Internet or the processor is that the former has a regular structure.

Figure 3 shows the typical layout of memory. It consists of an array of cells that are implemented as transistors. Each cell can hold either a zero or a one. Each horizontal collection of cells is known as a word and the depth of the array is, not unsurprisingly, known as the depth.

A circuit known as an address bus is connected to the array. Each word in the array has a unique identity known as its address. When data from memory or a program instruction are required by the processor of the computer, a signal is sent along the bus; this instructs the memory unit to decode the address and make the data or program instruction available at the specified location available to the processor.

There are a variety of memory devices available. Mask-programmed ROMs have their data or programs placed in them when they are fabricated and cannot be changed. Programmable Read Only Memories, commonly known as PROMs, are fabricated

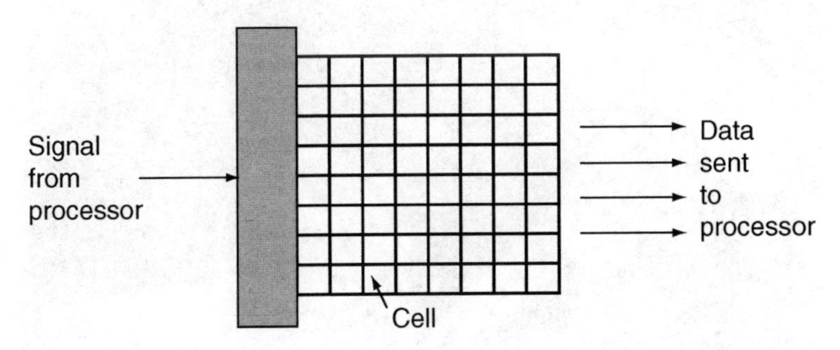

3. Computer memory

in such a way that they are blank and can then be programmed using an electronic device. However, since they are read-only this can only happen once – they cannot be reprogrammed.

Erasable Programmable Read-Only Memory, more commonly known as EPROM, goes one step further than PROMs in that it can be erased and then reprogrammed. There is confusion about EPROMs: since they can be reprogrammed, there is the impression that their contents can be changed by overwriting when they are in the computer. Overwriting can only be achieved by a special-purpose piece of equipment in which the EPROM device is inserted.

A major step forward that bridged the worlds of read-only memory and writable memory was the development of Electrically Erasable Programmable Read Only Memory, often known as EEPROM. This is a form of EPROM that can be erased while it forms part of a computer. A technology known as flash memory has been developed using the ideas and technologies associated with EEPROM.

Flash technology is employed in applications where a large amount of non-erasable memory is required. For example, it is used in the USB memory sticks that are used to transfer data from one computer to another, or as backup storage in case a computer

4. A hard disk unit

fails. Other applications that use flash technology include personal digital assistants, laptop computers, digital audio players, digital cameras, and mobile phones. A new-generation small laptop computer known as a netbook employs flash memory for the storage of programs and files of data.

Writable memory is implemented using a number of technologies. Dynamic Random Access Memory (DRAM) is the most common memory used within computers. It is implemented in such a way that it can potentially lose its data over a short period of time. Because of this, the contents of DRAM cells are read and written to continually in order to recharge its data.

Static Random Access Memory (SRAM) is a technology that does not require refreshing to be applied to its cells unless a program alters the cell or the power is removed from the computer it is contained in. It is faster but more expensive than DRAM.

File-storage technology

The technologies that I have described so far are normally used for relatively small amounts of data (it is a remarkable consequence of the advances in miniaturization and fabrication that I can refer to 8 Gb as 'relatively small'). For larger quantities of data and programs, a different, much slower technology is employed. It is known as hard disk technology.

In essence, a hard disk unit consists of one or more circular metallic disks which can be magnetized. Each disk has a very large number of magnetizable areas which can either represent zero or one depending on the magnetization. The disks are rotated at speed. The unit also contains an arm or a number of arms that can move laterally and which can sense the magnetic patterns on the disk. The inside of a hard disk unit is shown in Figure 4; here, the arm is clearly visible close to the edge of the disk.

When a processor requires some data that is stored on a hard disk, say a word processor file, then it issues an instruction to find the file. The operating system – the software that controls the computer – will know where the file starts and ends and will send a message to the hard disk to read the data. The arm will move laterally until it is over the start position of the file and when the revolving disk passes under the arm the magnetic pattern that represents the data held in the file is read by it.

Accessing data on a hard disk is a mechanical process and usually takes a small number of milliseconds to carry out. Compared with the electronic speeds of the computer itself – normally measured in fractions of a microsecond – this is incredibly slow.

Because disk access is slow, systems designers try to minimize the amount of access required to files. One technique that has been particularly effective is known as caching. It is, for example, used in web servers. Such servers store pages that are sent to browsers for display. If you look at the pattern of access to the web pages associated with a web site, you will find that some pages are retrieved very frequently – for example the home page – and some pages accessed very little. Caching involves placing the frequently accessed pages in some fast storage medium such as flash memory and keeping the remainder on a hard disk.

Another way of overcoming the slow access times to hard disks is to replace them completely with electronic storage media such as flash memory. Currently, such technologies do not approach the storage capacity and cost of hard disk units: typically flash memory can be bought up to 64 Gbytes and for the same price you can buy a hard disk of 500 Gbytes. However, for some types of computer which have modest storage requirements electronic memory for the bulk storage of files is now feasible. For example, some computers feature electronic storage media bulk storage rather than a hard disk. Such computers contain low power processors and the use of this memory compensates for this.

Future technologies

The last 30 years have seen amazing progress in miniaturization. There are, however, some performance and size plateaus which will be reached comparatively soon. For example, as you pack more and more components onto a silicon chip, they become more error-prone due to random signals being generated; another problem is that a phenomenon known as sub-atomic erosion kicks in, destroying the structures in the silicon. There are also design problems that occur with highly miniaturized circuits.

In the final chapter, I shall look at two radical ideas that will, if successful, transform computing. These are the use of biological materials for circuits and the deployment of quantum physics ideas in the development of processors; these require very long-term research. There are, however, other low-level technologies and materials that are currently being researched and developed for computer circuits. They include the use of optical lithography to produce components that are faster, the use of optical connections between components, the replacement of silicon by carbon, and the use of superconducting materials for the circuits within a computer.

Chapter 3
The ubiquitous computer

Four examples of ingenuity

A couple of years ago, I parked my car early in Milton Keynes, and as I was waiting to use the automatic ticketing machine, I noticed that the previous customer was a parking attendant. I made some remark about the expenses of his job being rather heavy if he has to park in Milton Keynes and whether he got an employee discount; he told me that his reason for getting a parking ticket was not to stick it on the windscreen of his car but as evidence that he had started work at the designated time – he was effectively using the machine as a time clock.

In *The Spy in the Coffee Machine*, Nigel Shadbolt and Kieron O'Hara describe an innovative use of the computer in Japan. The Japanese have a major problem with age drift: the birth rate of the country is one of the lowest in the world and immigration is discouraged. As a consequence, the population of Japan is rapidly ageing and there is an increased need to look after the elderly. One way that this is being achieved is via computer-based devices. One of these is the iPot. This is a kettle that keeps coffee or soup warm all day. Whenever the iPot is used a message is sent to a server and, twice a day, a report on the frequency of use is sent to a relative or carer either by a mobile phone text message or via email to satisfy them that those who are looked after are in good health.

The Japanese company NEC has developed a new type of spectacle that can project messages onto the retina of a user. There are a number of uses for such a device: two examples are to help staff who use computers in call centres and to provide a running translation when a conversation occurs in two different languages. (As you will see in Chapter 7, computer translation of languages has progressed to the point where quite decent results can be obtained, so this is not really an example of a science-fiction application.)

The Dutch firm Royal Dirkzwager has developed a system that tracks the movement of ships around the globe in real-time. It uses a satellite-based technology to track the position of ships on a minute by minute basis. It allows ships to be directed to ports that have space, reduces the time that a ship has to wait to berth, and reduces the amount of fuel that a ship requires.

All these examples have arisen from ingenuity allied with the miniaturization and the reduction of fabrication costs of the computer that I detailed in Chapter 2. They are just a small snapshot of the way that, unless you are a hermit, you will encounter computers in a variety of forms during a single day.

When writing this book, I took short walks of around forty minutes around my village: I find that it clears my head and gives me time to think. One of the walks is along quite a lonely track in the country around my house. During August, I walked along the track trying to think of circumstances when I was not in contact with a computer; I was beginning to think that this walk was one of the few examples. It wasn't: as I passed a gap in a hedge I noticed a combine harvester working in the distance. Such harvesters use sophisticated computer control to regulate the forward speed of the harvester and to keep the rotation speed of the threshing drum constant: this is quite a tricky thing to do and researchers in electrical engineering departments are still trying to develop techniques to optimize the threshing efficiency. Perhaps if I lived at the South Pole? Even there it would be difficult to get away from computers: scientists use computer tags to track the movement of penguins.

What is clear is that everywhere we go, we come into contact with computers and, increasingly, they have altered our lives. For example, a recent study by researchers at the University of California, San Diego, has estimated the amount of information delivered to us every day is equivalent to something like 100,000 words of text. In this chapter I shall be looking at how this information overload affects scientists, examining some positive aspects of computer ubiquity in health-care, and signposting some problems associated with privacy and confidentiality.

Computer ubiquity

There are a number of trends which have liberated the computer from the PC. The first is the increasing miniaturization of electrical components – not just hardware processors and memory, but also communication circuits and circuits used for signal monitoring. The second is the growth of technologies that enable wireless communication between computers. The third, and one that is often overlooked, is the increase in the ruggedness of electronic circuits: I have dropped my mobile phone so many times and yet it still functions.

Ruggedness means that computers can be attached virtually anywhere and still function, even in the most extreme conditions; for example, climate change researchers have attached computer-based measuring instruments to cargo ships and oil tankers in order to measure the temperature of the ocean as they carry out their journeys – such computers are battered by waves, have to undergo major variations of temperature, and suffer from engine vibrations, but still function.

Computers are everywhere: in iPods, mobile phones, security systems, car navigation devices, ATMs, automotive-electronic circuits, and so on. This has three major implications. The first implication is that it gives rise to a new discipline of ambient informatics in which data are available anywhere and at any time.

The second implication is that since these data are generated by our normal interactions in the world, for example by visiting a shop which senses a computer-based device that we may be carrying or a device embedded in, say, a pair of spectacles, or by driving to some destination using an intelligent satellite navigation system, there are major implications in terms of privacy and security.

The third implication concerns the mode of interaction. I interact with my PC in an artificial way in which I am always aware that there is a form of interaction going on. Ubiquitous computing involves an interaction that, in a sense, is natural because it is unobtrusive. Here's an example. One of the first applications of ubiquitous computing was associated with physical security and involved a computer being embedded within an identity badge known as an 'active badge'. The computer emits signals which are picked up by monitoring points in a building and provides information to staff about where colleagues or visitors are. Wearing an active badge is unobtrusive: you don't feel the wireless signals being emitted. It is less obtrusive than your heartbeat which, very occasionally, you are aware of.

In order to examine some of these implications, it is worth focusing on an ambient technology that is mature and cheap, and for which there are a number of working applications.

Radio Frequency Identification

Radio Frequency Identification, or RFID as it is more commonly known, involves the attachment of a small – they can be incredibly small (researchers at Bristol University have attached RFID devices to ants) – electronic device to an object or person. The RFID device, a tag, emits radio waves which can then be picked up by a computer equipped with a radio receiver. RFID tags were initially used in stock control where a retailer such as a supermarket attached them to items of inventory in order to track sales and stock levels.

These tags now retail for a few pence and their use has moved from the domain of stock control. For example, RFID tags are used in hospitals to track down equipment that may be required urgently; they can be attached to young babies or patients who suffer from dementia in order to discover their current location; they can be used as a repository for medical information: when a patient enters hospital they might be given a tag attached to a wristband so that when they, for example, attend some medical test the values of the test can be downloaded to the tag immediately; they can be used for timing participants in a sport such as track athletics; and they represent a potential technology that can be used within intelligent traffic management systems where they would be attached to cars.

Clearly, RFID tags have a major potential but there is a downside: that of privacy. There have been a number of boycotts of RFID-identified products over the fact that RFID tags affixed to products remain operational after the products have been purchased and can be used for surveillance and other purposes. A typical worry was expressed by California State Senator Debra Bowen at a privacy hearing:

> How would you like it if, for instance, one day you realized your underwear was reporting on your whereabouts?

Clearly, an RFID tag attached to the packaging of an item of food poses few privacy concerns: when the item is eaten the wrapping is normally disposed of. However, tags attached to devices such as mobile phones and MP3 players and to other items such as clothes, in conjunction with wireless readers situated in shopping malls, toll roads and other public places, provide the technological infrastructure for a surveillance society.

However, such tags could be very useful when an item is taken into a repair shop. One suggestion that addresses this and privacy concerns is to have a tag that has a section that can be ripped off when the item is bought and which contains the wireless transmitter part of the tag, leaving basic data that could only be read by a hand-held reader that would only be effective a few centimetres away.

Privacy concerns have also been expressed over the potential use of RFID tags that can be inserted into the human body. A company known as Verichip developed an RFID tag that can be surgically implanted into a human. The chip was approved by the American Food and Drug Administration. Clearly, there are some applications of this type of tag, for example carrying out continual monitoring of vital functions such as blood pressure; however, its wide-scale use raises major issues about privacy.

RFID represents the most advanced deployment of ubiquitous computing ideas. However, there are many others which are almost as advanced or over the horizon; it is worth looking at some of them.

Health

One of the major trends in this century has been the increasing amount of integration that has occurred, with the computer carrying out the role of a data processor and synchronizer between the hardware components. The best example of this is the iPhone, which functions as a mobile phone, personal organizer, MP3 player, and Internet access device.

In an article in the *New York Times* (5 November 2009), the columnist David Pogue describes how he was asked to speak at the TED Med conference for 18 minutes on medical applications for the iPhone. TED conferences are organized by a non-profit-making organization which has as its main aim the spreading of ideas (their web site, http://www.ted.com/, is fantastic and contains videos of most of the presentations).

Pogue was worried that he would not find many applications to talk about and he would not fill the 18 minutes. The problem that he did encounter was that he found far too many: over 7,000 applications – just for the iPhone. For the normal user, these included Uhear, an application which enabled the iPhone to test someone's hearing; ProLoQuo2Go, a speech synthesizer for people with speech difficulties that enables the user to touch phrases, icons, and words on the screen of the iPhone and then speaks the text that they have identified; and Retina, an application that allows a colour-blind user to point their iPhone at some coloured item which would then identify the colour.

Pogue also discovered applications for medical staff. These included Anatomy Lab, a virtual cadaver targeted at medical students that allows the user to explore human anatomy; Epocrates, an electronic encyclopaedia which, for example, would warn a doctor about the side effects that two prescribed drugs might have; and AirStrip OB, which, for example, enables an obstetrician to monitor a pregnant mother's vital signs remotely.

Ubiquitous computing also has major applications in the care of the elderly. At an international conference on the use of ubiquitous technology for assistive care, four researchers from the computer science department at the University of Texas, described the design of a simple and cheap wireless network that could be deployed in care homes or the homes of the elderly. This is similar to the use of the iPot detailed earlier that monitors the state of elderly Japanese in their homes.

The network they described would support the sort of monitoring that could easily implement a wide variety of applications, ranging from detecting movement and issuing a message to a remote monitoring station when no movement was detected after a certain time period, to the monitoring of vital signs such as heart rate using RFID technologies. One of the major expansion areas in consumer electronics over the next decade will be that of wireless-based home entertainment where all the wired connections would be replaced by wireless connections. The sort of network described by the researchers could be easily piggy-backed on top of these local networks.

The global panopticon

Computers can be placed in a variety of places: in satellites, in remote telescopes, in temperature measuring systems and even attached to cargo ships. The last decade has seen a major increase in the amount of data that has become available to researchers in areas such as geology, climatology, and physical geography.

There are a large number of examples of computers embedded within hardware that generates huge quantities of data. The Australian Square Kilometre Array of telescopes will generate billions of items of data, the Pan-STARRS array of celestial telescopes will generate several petabytes of data per day, and the technology of gene sequencing is quickly advancing to the point that billions of items of DNA data can be generated in just a few days.

The output from such computers is hugely valuable and has, for example, transformed our study of climate change. However, while the computer has provided the gift of such data, it has posed major problems. The data provided by equipment such as nuclear colliders, gene sequencers, and radio telescopes are of value to the whole scientific community – not just those who have carried out experiments using this hardware. Unfortunately, there are very few standards for the storage and presentation of such data.

There are a small number of scientific data libraries in the world. A good example is the San Diego Supercomputer Center at the University of California, San Diego. Here a large amount of data (currently many petabytes) including bio-informatics and water-resource experiments is stored in such a way that other researchers can access it. Another example is the Australian National Data Service (ANDS), which offers a registration service for data. ANDS does not store data, it stores information about data stored elsewhere: the web site where it can be accessed, the nature of the data, and who was responsible for the generation of the data are just three items that can be accessed via the ANDS computers.

It is not just the amount of data that is increasing, but also the research literature. Medical research is a good example of the explosion. In 1970, there were approximately 200,000 research articles catalogued; this had risen to close on 800,000 by 2009.

During the writing of an early draft of this chapter a major security incident affected the Climatic Research Institute (CRU) of the University of East Anglia. CRU is one of the premier climate research institutes in the world and was responsible for a number of databases including one that contained data from temperature measuring stations around the world.

In October 2009, hackers took a massive file of emails, documents, and program descriptions from one of the computers at the CRU. This theft occurred a few weeks before a major climate change conference in Copenhagen and the contents of the files that were taken caused a blizzard of blog postings, articles, and emails. At its height, I counted over 30 million references to 'Climategate', as it came to be known.

Climategate had a number of dimensions. There were arguments about whether scientists at CRU had massaged data, whether the behaviour of the scientists at CRU was ethical with respect to other scientists who did not share their views about global warming, and whether the United Kingdom's Freedom of Information Act had been violated.

What was clear, however, from this incident was the diverse nature of climate data throughout the world, the lack of proper cataloguing at the various repositories of the data, and the lack of visible program code that manipulated the data.

If the scientific world is to make any headway in terms of fully harnessing the data-gathering and data-processing of computers, then a major shift is required comparable to the political, intellectual, and scientific efforts that created the multi-disciplinary teams that came together for the Manhattan Project or the Bletchley code-breaking projects that had a major effect on the outcome of the Second World War.

Foremost amongst those pressing for the use of the Internet as an open repository of scientific data was Jim Gray, a researcher at Microsoft. He was a computer science researcher who pressed hard for the scientific community to come to terms with the change that computers are bringing to their research. He envisaged a fourth paradigm which would sit alongside the existing three scientific paradigms of empirical observation, analytic processing of the data (usually using statistical methods), and simulation or modelling based on the analysis whereby theories associated with the phenomenon that generated the data are created or modified. Gray's fourth paradigm has a number of components which give rise to both technical and political challenges.

The first is that of the curation of experimental data. This involves the development of Internet sites dedicated to the storage of data, metadata, and any computer programs that are associated with the data, for example a program that carried out some manipulation of the data before it was deposited in a database. This is a major challenge: it does not just involve storing a large quantity of data on a web site, but also involves the specification of metadata – data that describe what each individual collection of data means – for example, the fact that a series of readings was taken from a particular temperature-measuring station between two dates. It also involves storing the details of how the data were changed and the processes and programs used to effect the change.

The second aspect of Gray's fourth paradigm has to do with the explosion of research publications. If you look at the structure of a research paper, you will see references to other papers, references to data sets, and to computer programs. Increasingly, the amount of intellectual effort required to read and digest such papers is approaching overload.

Researchers are starting to address this problem. One way to help the reader is by a form of enhancement whereby supporting materials are added to a raw article. An example of a tool that does this is EMBL Germany's *Reflect* tool. It tags gene, protein, and small molecule names within a research paper, with the tags being hyperlinked to the relevant sequence, structure, or interaction databases that hold the bio-informatic data. So, if a reader wants to cross-reference a gene sequence, all they have to do is to click a link within the paper.

A third component of Gray's fourth paradigm is that of data visualization. When you have a large quantity of data, you will have a lesser but still large quantity of output results after those data have been processed. Increasingly, the size of such output is defeating conventional ways of display such as two-dimensional graphs. Researchers are now striving to develop novel – often three-dimensional – ways of rendering the output such that the viewer can, for example, discern patterns.

The fourth component of Gray's fourth paradigm is the integration of all the elements that make up a computer-based scientific experiment. When you carry out an experiment, you start with a conceptual view of what the experiment will do, you then transform this into a concrete procedure involving the gathering of data, you then process the data using some computer program which displays the results, and, in the final step, you publish the results in some academic journal or conference. What is needed is a chronicle which describes every step in this process, together with a link to all the relevant documents, data, and program codes used in the experiment. This is, without a doubt, the greatest challenge facing those scientists who involve a computer in their experimentation.

Chapter 4
The global computer

Introduction

In this section, I hope I will convince you that to think of the computer as the box that resides on a desk or as a lump of silicon that is used in control and monitoring applications such as those found in avionics applications and chemical plant monitoring is restrictive. I hope that I can convince you that by connecting computers – their processors and their memory – together we can, in fact, create larger computers; the ultimate instantiation of this being the Internet.

The key to the development of the global computer is its processor: an electronic circuit that reads, decodes, and executes the instructions in a computer program and carries out the intentions of the programmer. The speed of computers has increased by orders of magnitude over the last 50 years. As soon as computer technology advances in terms of performance (processor speed, size of memory, and memory-access speed), new applications come along which require even more speed, larger memory, or faster access to the memory, or there is a demand for an improvement in a current application such as weather forecasting, where hardware advances have made predictions more accurate and enabled the forecasters to reach out further into the future.

Wicked problems

Before looking at how we create more and more powerful computers, it is worth looking at some of the major problems they have to solve – so called 'wicked problems' that require huge computational resources for their solution. The world is full of problems that are wicked; they require huge amounts of computer resource and human ingenuity to solve. One of these was the Human Genome Project. This project discovered genetic sequences. If the sequences obtained were to be stored in books, then approximately 3,300 large books would be needed to store the complete information. The computational resources required to search the gene database to look for genes that predispose someone to a particular disease or condition are massive and require supercomputers.

The follow-on projects are progressing slowly since the computational demands are huge and can only be satisfied by the supercomputers that I describe later in this chapter, but it is progressing. There are, however, a class of problems that are incapable of being solved exactly by the computer. They are known as 'NP-hard problems'.

One of the surprising features of many of these problems is that they are simple to describe. Here's an example known as the set partition problem. It involves deciding whether partitioning a set of numbers into two subsets such that the sum of the numbers in each set are equal to each other can be achieved. For example, the set

(1, 3, 13, 8, 6, 11, 4, 17, 12, 9)

can be partitioned into the two sets

(13, 8, 9, 12)

and

(4, 17, 6, 1, 3, 11)

each of which adds up to 42. This looks easy, and it is – for small sets.

However, for much larger sets, for example sets containing tens of thousands of numbers, the time taken to carry out the processing required to discover whether the sets can be split is prohibitive; it rapidly reaches the point where, with even the most powerful computers that have been constructed, the time would exceed the known life of the universe.

Such problems are not academic: they often arise from practical applications; for example, NP-hard problems arise in the design of VLSI circuits, the analysis of genetic sequences, and in avionics design. One of the most famous NP-hard problems is known as the travelling salesman problem and arose from a task associated with the design of computer hardware. Here the aim is, given a series of cities and the distances between them, to develop a route for someone (the travelling salesman) that takes them to each city at the same time as minimizing the route and hence the amount of petrol used.

In practice, the vast majority of NP-hard problems do not require an exact solution – a solution close to the exact solution would do. For example, there is an NP-hard problem known as the bin-packing problem where the computer is given a number of containers and a series of boxes and the aim is to minimize the amount of slack space in the containers. For this problem, it is possible to get within 99.5% of an optimal solution.

Because of this, a large amount of research connected with NP-hardness concerns what are known as approximate algorithms. These, as the name suggests, are descriptions of computer programs that produce approximate but good enough solutions.

Solving wicked problems by software

Recently software researchers have harnessed a number of ideas in biology to improve the capabilities of programs that try to produce approximate solutions. One of these is genetic programming. Here a set of candidate programs are generated to solve the problem and then run. The top programs in terms of efficiency are then collected together as a new generation of programs and combined together to create a further generation of programs. This generation is then run and further mating occurs until a suitable program that works efficiently emerges.

The term 'genetic programming' comes from the fact that the process of generating more and more efficient programs mimics the Darwinian process of evolution (genetic programming is often known as evolutionary programming). In common with the other techniques I detail in this section, it requires considerable computer resources.

There are a number of other techniques that mimic biological processes in order to solve wicked problems. Swarm computing is based on a model that draws on behaviour exhibited by birds within a flock, insects cooperating with each other over a task such as food gathering, or fish swimming in a shoal. Such behaviour can be modelled in very simple ways and this simplicity has been transferred to a number of optimization programs.

There are also computer programs that mimic the behaviour of social insects such as ants, for example, processing data in the same way that ants forage for food or dispose of their dead. The problems that such ant-colony programs solve are concerned with an area known as routing; here the underlying data can be modelled as a series of points connected by lines, for example the layout of connections on a VLSI-fabricated chip.

There are also other ways of programming wicked problems. There is a controversial area of computer science known as

5. **The CRAY XM-P48**

artificial intelligence – controversial because of the over-claiming that its proponents have made over the last thirty years. Researchers in this area belong to two camps: those who try to use the computer to gain an understanding of how humans carry out processes such as reasoning and those who just want to develop intelligent software artefacts whose performance matches that of humans – irrespective of whether the software bears any resemblance to human processes.

The artefact builders have developed a number of techniques that have been successful in the real world. Almost certainly the best known product of this approach to artificial intelligence is 'Deep Blue', a chess-playing program that defeated the world chess champion Gary Kasparov in 1997.

This program relied on the massive calculating power of a supercomputer and did not base much of its power on studies of how human chess players behaved. There are however a number of artificial intelligence programs that attempt to overcome wickedness by a combination of brute computational force and human heuristics. One type of program is known as an expert system. It attempts to carry out tasks that humans carry out such as diagnosing illnesses or finding hardware faults in electronic systems. Such expert systems rely not only on the computer's power but also on an encoding of some of the heuristics used by human experts in the domain they work in.

Even given the advances in software technology that we have seen over the last twenty years there is still a need for powerful computers: the software technology goes some distance to solving big problems – but not far enough – and there is the natural tendency to move on and attack bigger problems and more complicated problems. So, the last 40 years has seen major advances in supercomputing.

Supercomputers

The first computers had a single hardware processor that executed individual instructions. It was not too long before researchers started thinking about computers that had more than one processor. The simple theory here was that if a computer had n processors then it would be n times faster. Before looking at the topic of supercomputers, it is worth debunking this notion.

If you look at many classes of problems for which you feel supercomputers could have been used, you see that a strictly linear increase in performance is not achieved. If a problem that is solved by a single computer is solved in 20 minutes, then you will find a dual processor computer solving it in perhaps 11 minutes. A 3-processor computer may solve it in 9 minutes, and a 4-processor computer in 8 minutes. There is a law of diminishing returns; often, there comes a point when adding a processor slows down the computation. What happens is that each processor needs to communicate with the others, for example passing on the result of a computation; this communicational overhead becomes bigger and bigger as you add processors to the point when it dominates the amount of useful work that is done.

The sort of problems where they are effective is where a problem can be split up into sub-problems that can be solved almost independently by each processor with little communication.

The history of supercomputing can be split into two phases: the 1970s and the 1980s and the years after these two decades. Before looking at the history, it will be instructive to see how speeds of supercomputers have increased over the last 70 years.

The first real supercomputers were developed by two companies, CDC and Cray. The most successful designs were based on a vector architecture. This is based on a processor that is capable of carrying out a number of instructions on data simultaneously, for example adding a thousand numbers together at the same time. The computers made by the Cray Corporation were the iconic supercomputers. Figure 5 shows a CRAY XM-P48, an example of which was situated at the Organisation Européenne pour la Recherche Nucléaire (CERN). It resembles the sort of furniture designed for the waiting room of a 1980s advertising company. However, when it was delivered to research labs around the world it was the fastest computer in existence: in 1982 it was a state of the art computer it had a theoretical top speed of 800 MFLOPS from both its processors (an MFLOP is a million instructions that carry out some arithmetic operation such as adding two numbers together).

Supercomputers were delivered to a variety of customers including CERN, the Los Alamos National Laboratory in the USA, the Boeing Corporation, the British Meteorological Office, the National Aerospace Laboratory in Japan, the US National Nuclear Security Administration, and the US Department of Energy Laboratory at Oak Ridge.

The customers for such behemoths betray the sort of applications they were used for: nuclear experiment simulations, weather forecasting, simulating the day-to-day processes that occur in a nuclear reactor, and the aerodynamic design of large planes. The key similarities between each of these applications are the volume of computation that needs to be carried out and the fact that most of the computations involve number crunching.

'Vector architecture computers' dominated supercomputing until the 1990s when mass-produced processors started becoming so cheap that it became feasible to connect them together rather than design special-purpose chips. One of the fastest computers in the world is the Cray XT5 Jaguar system which has been installed at the National Center for Computational Sciences in the USA. It has around 19,000 computers and 224,000 processing elements, based on standard hardware processors rather than bespoke designed processors.

Even small-scale research establishments can now get in on the act, often by developing their own version of a supercomputer out of commercial hardware processors. These computers are known as Beowulf clusters. Such computers are based on readily available processors such as the ones you find in your home PC, the LINUX operating system – a free operating system often used by researchers for scientific computation – and other open-source software.

The power of Beowulf computers is huge and they are cheap to make: the Microwulf cluster that was developed by computer science professor Joel Adams and student Tim Brom weighed just 31 pounds (small enough to fit in a suitcase) and had speeds of up to 26 Gflops. The cost of this computer in early 2007 was about $2,500.

In 2009, Intel, the leading chip maker in the world, announced a new processing chip containing 48 separate processors on a single chip. Such chips used in a Beowulf computer would bring real supercomputing power to the desktop.

The Internet as a computer

This chapter started out describing supercomputers: machines that carried out huge numbers of computations. Such computers can employ thousands of processors. What is surprising is that there is a much larger computer and anyone who uses a desktop or laptop computer is part of it. It's the Internet.

The Internet has been described as a network or, more accurately, as a network of networks. Let's see how this works out. When I order an item from an online bookseller such as Amazon, I click on a series of links which identify me and which identify what books I want to buy. Each time I click, a message is sent to a computer at Amazon known as a web server. The computer discovers which page I want and sends it back to me, for example a page containing the final order that I make. I then click on another link – usually the link that informs the bookseller that I have finished my order.

As part of the interaction between my browser and the Amazon web server other computers are used. First, there are the computers that transmit the messages from my computer to the web server – they do not go directly but are split into packets each of which may go via a set of completely different computers.

Second, there is the collection of computers that are known as the Domain Name System (DNS). The DNS is a hugely important part of the Internet. When you type in a symbolic name for a web site, such as http://www.open.ac.uk, it is the DNS that discovers where on the Internet this web site is situated. Without this collection of computers, the Internet would be unable to function.

There are other computers that are involved in the sales process. A book retailer will have a number of computers known as database servers. These contain large collections of data; in the case of the online bookseller, they would hold data such as which books are on sale, their price, and the number in stock. They would also contain data such as past sales to individuals and marketing information, such as the amount of sales for past published books. When the web page that contains details about availability is presented to a customer, then a database server is involved.

There will also be computers at the bookseller's warehouse. One application of such computers is to provide what is known as a picking list: this is a list of books that have been bought within a certain time period. The computer will sort these books into a list that details each book and its position in the warehouse and will organize the list in such a way that the amount of travelling needed by warehouse staff is minimized. There will also be a computer that is used by warehouse staff to inform the database server that books have been picked and need to be marked as being taken from the shelves.

There will also be computers that carry out financial functions, sending money transfers to book suppliers for the books that have been bought from them, and sending debits to credit card companies for the purchases made by the customer. So, what this example shows us is that the Internet functions as a series of computers – or more accurately computer processors – carrying out some task such as buying a book. Conceptually, there is little difference between these computers and the supercomputer, the only difference is in the details: for a supercomputer the communication between processors is via some internal electronic circuit, while for a collection of computers working together on the Internet the communication is via external circuits used for that network.

The idea that network technologies used in the Internet can be used to create a sort of supercomputer has been embedded in something known as the grid and the resulting computer known as the grid computer.

The grid

So, what is grid computing? Well a computer grid is a collection of conventional computers that are linked together using Internet technology and usually connected by high speed communication circuits. There are two ways of looking at the grid: that it is the continuation of mainstream supercomputing ideas – this is the view taken by researchers – or that it is a new way of optimizing the use of computers in an enterprise. You will remember that computer processors have a large amount of slack when they are used: they can be idle for as much as 95% of the time. The vendors of commercial grid software make the point that buying their product will reduce an enterprise's hardware costs by a significant amount.

Grids can be formal or informal; the former is usually supported by commercial software which maintains the grid, allowing file sharing and processor sharing, the latter is a loose confederation of computers that carry out some large task. A good example of an informal network is folding@home. This is a network that is coordinated by the Stanford University Chemistry department. Its aim is to carry out much of the number crunching associated with protein folding; this is work associated with the Human Genome Project and which attempts to find cures for serious diseases such as Parkinson's Disease and Cystic Fibrosis. The network has hundreds of thousands of computers connected to it with a combined speed approaching 4 pFLOPS.

Grid computing represents a subtle shift from the idea of a supercomputer carrying out a massive number of calculations in order to solve a wicked problem towards commercial applications outside the realm of number crunching. In application terms, it has influenced the idea known as cloud computing which regards the Internet as a central utility for all applications – not just number crunching – and which threatens to overturn the current way that computers are used commercially. I will discuss this further in Chapter 7.

Afterword

What this chapter has shown is that to think of a computer as just a box sitting on a desk or as a piece of silicon-based circuitry in something like a DVD player or a set of traffic lights is too limited: that regarding the Internet as a large computer raises a host of interesting questions which I will address in Chapter 7, there I discuss an evolving model of commercial computer use that takes much of the processing and data away from individual computers and delegates it to powerful servers maintained by commercial enterprises.

This is also a theme I shall look at in the final chapter, where I will examine the work of Jonathan Zittrain and Nicholas Carr. Zittrain has posited that the freewheeling growth of the Internet has enabled a new age of creativity amongst computer users, but at a cost in terms of problems such as security. His book *The Future of the Internet* describes a possible scenario where commercial pressures close down much of what the ordinary computer user can do to the point where the home computer is relegated to something like the dumb terminal of the 1960s and when a golden age of computational creativity ends.

Carr uses the analogy of the Internet as a computer but concentrates on an industrial viewpoint. He posits a future where computing power becomes a utility in the same way that electrical power becomes a utility and where the role of the computer – at least for the home owner – is reduced to that of accessing the Internet. I shall return to this theme in the final chapter.

Chapter 5
The insecure computer

Introduction

During the 2009 Iranian election protests, foreign activists disabled the computers acting as web servers and belonging to the Iranian government by carrying out what is known as a denial of service attack. The attack took place as part of the protests against what many saw as a corrupt election result. The activists flooded the servers with hundreds of thousands of requests for web pages to the point where the processors of the web servers were overwhelmed by the amount of data being sent. This effectively closed them down.

In 1999, a computer virus known as Melissa was released into the Internet. What the virus did was to infect the Outlook email program which formed part of the Windows operating system. The virus was spread using email. If someone received an email that contained the virus as an attachment and then clicked on the attachment, their computer would be infected. Once a computer was infected by the virus, it accessed the contact list of Outlook and emailed the first 50 contacts on this list and sent the virus to them. This was a particularly pernicious virus, not just because it spread rapidly, but because it also had the potential to modify word-processed documents on an infected computer so that they also spread the virus.

In 2007, the British government reported that child benefit data for over 25 million people entrusted to Her Majesty's Revenue and Customs department had been lost. The details were stored on two CDs which were sent in the post to another department.

These are just three examples of computer insecurity. The first two involved technical failings while the last was a managerial failing. The aim of this chapter is to look at the threats to computers and how they can be countered. It will also look at some of the entwined issues of privacy.

Viruses and malware

A virus is a computer program that is introduced into a computer or a network of computers illegally; once introduced it carries out some malicious act. Typical malicious acts include: deleting files; in the case of the Melissa virus, emailing documents out to other Internet users; monitoring the keystrokes carried out by a user in order to discover important information such as passwords and banking information; and scrambling a file so that it becomes unreadable and then asking for a ransom which, when the victim pays, will result in an email being sent to the victim with instructions on how to unscramble the file. Viruses are a subset – albeit a large subset – of a collection of software known as malware.

There are three main ways for a virus to be introduced into a computer system. The first is as an attachment, for example a user may be sent an email which contains a message about the surprising behaviour of a celebrity with instructions to click on the attachment in order to see photographic evidence of the behaviour; as soon as the recipient of the email clicks on the attachment the virus will have taken up residence on the computer.

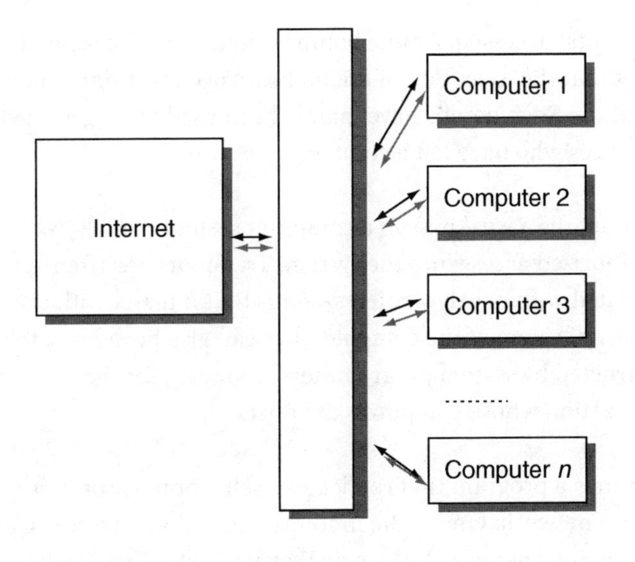

6. The architecture of a firewall

The second way that a computer catches a virus is by using insecure software. There is a host of free software available on the Internet; most of it is useful (I use an excellent program for backing up my files), however, some software will contain viruses. As soon as the software that contains the virus is installed, the virus is installed as well. The carrier of the virus is often known as a 'Trojan horse'.

The third way is by not employing or not properly employing a program known as a firewall. This is a barrier that prevents unwanted connections into a computer. It continually monitors these connections and rejects any that are deemed to be harmful.

There are a number of other types of malware apart from viruses and Trojan horses. A logic bomb is a program that executes within a computer when a particular condition occurs, for example when a file is first accessed. A time bomb is similar to a logic bomb except that the execution of the malware occurs at some time in the future. Such bombs have mainly been used by disgruntled employees who have left a company.

A trapdoor is a weakness in a computer system that allows unauthorized access into the system. Trapdoors are often associated with poorly written systems that interact with the operating system of the computer, but can also be deliberately constructed by system programmers who work for the organization whose computers are affected.

A worm is a program that replicates itself a number of times across a network. One of the first intrusions that were associated with the Internet was the Morris Worm which affected a large number of computers in the emerging Internet.

A rabbit is similar to a worm except that it replicates itself on one computer rather than a network and effectively shuts down the computer by hogging all its resources – main memory, file storage memory, or even the hardware processor can be affected.

Viruses can be hidden in a number of locations, for example they can be integrated with programs: they can surround a program and intercept any requests for the program to be carried out, they can be appended to the end of a program, or they can be embedded within a program. They can also be embedded within a document or an image and then executed when the document or image is opened.

An important technology that is used to detect, isolate, and delete viruses is the virus scanner. There are a number of commercial and free virus scanners that are available via companies such as Norton, Kaspersky, and AVG. A virus is just a program – albeit a program that can cause major destructive effects – and like all programs has a signature. This is the pattern of zeroes and ones that is unique to the virus. A virus scanner will, when executed, process the hard disk and memory of a computer looking for virus signatures. In order to do this, it relies on a database of signatures of known viruses.

When you purchase a virus scanner, you also purchase a database of virus signatures and a program that periodically updates the database as new viruses are discovered and examined by the company that has sold you the scanner.

Pfleeger and Pfleeger in their excellent book on computer security, which is cited at the end of this book, describe the steps that you need to take in order to protect yourself from viruses: use virus scanners; use only commercial software that has been purchased from reputable vendors; test all new software that you are suspicious of on an isolated computer; if you receive an email with an attachment then only open it if you know that it is safe (one check is to type some of the key words of the email into a search engine such as Google: you may possibly find that someone has flagged the email as hazardous); and always backup your files, for example moving them to a cheap hard disk that is normally kept separate from your computer.

Another technology – already briefly referred to – that is highly effective against virus and other attacks is the firewall. There are a variety of firewall architectures. One architecture is shown in Figure 6. What a firewall does is to intercept any traffic from outside a protected network – usually from the Internet – and reject any traffic that can be dangerous for the network. There is also a firewall known as a personal firewall that protects individual computers; however, its function is similar to an industrial firewall: it's just that it has a smaller set of functions. Such firewalls are normally marketed for home use and are sold as an integrated package that contains a virus scanner and other security software.

Typical actions a firewall carries out are: checking any attachments to see if they contain viruses, denying access to a computer or a network to messages sent from a computer that has been identified as the source of hazardous traffic, and denying facilities to external computers such as copying files.

Computer crime

There are a large number of crimes associated with the computer, including: fraud achieved by the manipulation of computer records; in some countries, spamming, the sending of unrequested emails usually selling some product or service, is illegal; accessing a computer system without permission; accessing a computer system in order to read, modify, and delete data without permission; stealing data or software; industrial espionage carried out via a computer; stealing the identity of someone by accessing their personal details; accessing financial data and stealing funds associated with that data; spreading viruses; and distributing child pornography.

A typical non-technical crime is where an employee has access to a financial system and issues a number of cheques payable to themselves and/or others. Such crimes have often involved figures of over a million dollars before the perpetrator was caught.

In 2008, a job vacancy was posted on a web site that contained adverts for translators. The job involved correcting texts which had been translated into English from another language. Applicants were asked to provide data such as that normally found on a CV, plus their bank account details. The 'company' that issued the advert was connected to a crime syndicate in the Ukraine. Anyone who applied would have their account siphoned of cash and then used for a money-laundering scam.

In 2009, a survey of American computers discovered that many millions of computers in the United States were 'infected' with fake security software. This is the result of a very sophisticated scam in which, when a computer user visited a particular site, they saw an alert in a pop-up window that told them that they had a virus infection and that free software could easily be downloaded and be used to detect and remove the virus along with many others. The software was, of course, a virus that was capable of discovering financial details of the user of the computer.

In 2006, three Russian hackers were jailed for eight years for carrying out an attack known as a denial of service attack on British bookmakers. In this type of attack, a computer is flooded with messages – sometime they are requests for web pages and sometimes they are email messages – with the result that the computer that is attacked is disabled and cannot respond to normal users' requests. The hackers targeted the web server of a major online bookmaker, who refused to pay a large ransom, their computer was blocked during the Breeders' Cup races, and the company lost more than £160,000 for each day of the attack.

An important point to make about computer crime is that many crimes can be carried out by computer users who have little technical knowledge. The examples above require varying degrees of technical skill with the money-siphoning example requiring the least. Typical non-technical crimes include: rummaging through large refuse bins for computer printouts containing sensitive information (a process known as dumpster diving), taking a photograph of important information about a computer system written on a white board using a mobile phone, and stealing a password and using that password to masquerade as the user of a computer system.

All these non-technical examples require, at most, the sort of computer knowledge that you gain by using a home computer for word processing or spreadsheet work. It is no surprise, then, given that computer crime can be carried out in a technical or non-technical way, that there is a variety of technical and non-technical defences against computer crime.

Technical defences against computer crime

One of the major defences against the large number of computer-related, technical crimes is cryptography. This is a technology used to hide and scramble data – data that may be stored in a computer or may be in transit between two computers connected by a network. Cryptography has been around since Roman times when Julius Caesar used it to send messages to his military commanders. The ciphers that are used in computing are quite complicated; however, a description of the Caesar cipher will give you a rough idea of how they work.

The Caesar cipher involves changing a text by replacing each letter in a message by the *n*th letter forward from it in the alphabet. For example, if *n* was two then the letter 'a' would be transformed to the letter 'c'; letters at the end of the alphabet would be transformed into letters at the beginning, for example the letter 'y' would be transformed to 'a'.

Caesar was here

would be transformed to

Ecguct ycu jgtg

In the Caesar cipher, the number *n* acted as a password. In cryptography circles, it is known as the key: it was used to determine the precise transformation that was applied to the text that was scrambled (the plain text) in order to produce the text that could not be read (the cipher text). All that would be required of the recipient of the message above would be to know the key (the number of places each letter was transformed forward). They would then be able to extract the message from the cipher text.

The Caesar cipher is very simple and can be easily cracked. Cracking modern ciphers is a complicated business that requires a knowledge of statistics to understand. It requires data about word frequency, letter frequency, and, ideally, some knowledge of the context of the email and of some possible words that could be contained in the message. It was knowledge of phrases that could appear in a message that enabled cryptographers in the Second World War to decipher German messages. For example, many of the messages would contain the words 'Heil Hitler'.

We have come a long way since the Caesar cipher; indeed, a long way since the Second World War. Computer-based ciphers are still based on substitutions such as the type found in the Caesar cipher. However, two factors have enabled ciphers to become very much stronger. The first is that transpositions are applied to the plain text as well as substitutions; a transposition being the movement of a character in the text to another position in the text. The second is that a huge number of substitutions and transpositions are carried out, normally via computer hardware or very efficient computer programs.

The type of cipher that I have described is known as a symmetric cipher; it is symmetric because the same key would be used for changing the plain text (encrypting it) as for transforming the cipher text back to the plain text (decrypting it). Symmetric encryption methods are very efficient and can be used to scramble large files or long messages being sent from one computer to another.

Unfortunately, symmetric techniques suffer from a major problem: if there are a number of individuals involved in a data transfer or in reading a file, each has to know the same key. This makes it a security nightmare. It is exacerbated by the fact that the more complex a key, the less likely it is for a message to be read; consequently, a key such as 'Darrel' is useless, while a key such as

55Tr3e343E4((!!dfg21AVD90kk - 9jk}

provides a high degree of security, the only problem being that it can be very difficult to remember and would need to be stored somewhere such as a wallet or within a drawer in a desk.

While there have been a number of attempts at making the key distribution process secure, for example by encoding keys on biometric smart cards and devising secure protocols for the distribution of a key or set of keys, key security for symmetric encryption and decryption still remains a problem.

One public solution to this problem was developed by three American computer science researchers, Whitfield Diffie, Ralph Merkle, and Martin Hellman, with a practical implementation of their ideas by three other researchers: Ronald Rivest, Adi Shamir, and Leonard Adleman. I use the word 'public' since declassified documents from the British government indicate that three British researchers, Clifford Cocks, Malcolm Williamson, and James Ellis, were also developing work in this area at GCHQ, the British government's top-secret communications headquarters in Cheltenham. Their work was carried out in the early 1970s.

The solution that was developed by these researchers was known as public key cryptography or asymmetric cryptography. At its heart, it relies on the inability of computers to solve difficult problems that require massive computational resources. The term 'asymmetric' best describes the technique, as it requires two keys, a public key and a private key, that are different.

Let us assume that two computer users *A* and *B* wish to communicate using asymmetric cryptography and each has a public key and a private key. The public key is published by each of the two users. If *A* wants to send an encrypted message to *B*, then she will use *B*'s public key to encrypt the message. When *B* receives the message, he will then use his private key to decrypt the message. Asymmetric cryptographic systems have a number of properties; a very important one is that the public key that is used to send the message cannot be used by someone who intercepts the message in order to decode it.

At a stroke, public key cryptography removed a major problem associated with symmetric cryptography: that of a large number of keys in existence some of which may be stored in an insecure way. However, a major problem with asymmetric cryptography is the fact that it is very inefficient (about 10,000 times slower than symmetric cryptography): while it can be used for short messages such as email texts, it is far too inefficient for sending gigabytes of data. However, as you will see later, when it is combined with symmetric cryptography, asymmetric cryptography provides very strong security.

One strong use is to provide a form of digital identity as a digital signature in which the mathematical properties of the key generation process are used to provide authentication that the person who purports to have sent a message is, in fact, that person.

A second use is in sending secure messages, for example between a bank and a customer. Here the identity of a user and the web server is mediated by a digital identity check. It is worth saying here that you should not rely solely on technology, but use common sense. For example a popular fraud is to send a user an email purporting to be from a bank asking them to check into a rogue site which masquerades as the bank site and extracts out account data that is used for siphoning funds. The Internet is full of villains who are experts in subverting the technology.

There are a number of technologies that are used to provide this security; almost all are based on a combination of symmetric and asymmetric cryptography. One very popular security scheme is known as the Secure Sockets Layer – normally shortened to SSL. It is based on the concept of a one-time pad. This gives an almost ideal cryptographic scheme. It requires the sender and receiver to have a document known as a pad which contains thousands of random characters. The sender takes the first n characters, say 50, and then encrypts the message using them as the key. The receiver, when they receive the message, takes the fifty characters from their pad and decrypts the message. Once these characters are used for the key, they are discarded and the next message is sent using the next n characters.

A major advantage of the one-time pad is that once a password is used it is discarded; another advantage is that any document can be used to generate a key; for example, a telephone directory or a novel could equally well be used. The main disadvantage of a one-time pad is that they require synchronization between the sender and receiver and a high degree of security for the printing, distribution, and storage of the keys.

SSL uses public key cryptography to communicate the randomly generated key between the sender and receiver of a message. This key is only used once for the data interchange that occurs and, hence, is an electronic analogue of a one-time pad. When each of the parties to the interchange has received the key, they encrypt and decrypt the data employing symmetric cryptography, with the generated key carrying out these processes. If the two parties want a further interchange of data, then another key is generated and the transfer proceeds as before.

Another technical defence against computer crime is the password. The theory behind passwords is simple: that the user of a computer remembers some set of characters and uses them to access a web site or a file – the password, in effect, acts as a proxy for the user's identity. Sometimes the password is used as an encryption key. Passwords offers major advantages; however, password schemes can be less secure because of human failings.

In general, computer users choose poor passwords. The best passwords and cryptographic keys are those which are long and which contain alphabetic, numeric, and special characters such as !. For example, the password

s22Akk;;!!9iJ66s0 – iKL69

is an excellent password, while 'John' is a poor one. In the 1990s, the online bank Egg discovered that its customers chose poor passwords. For example, 50% chose family members' names. Charles and Shari Pfleeger, in their excellent book *Security in Computing*, provide sensible advice on password choice: don't just use alphabetic characters, use long passwords, avoid names or words, choose an unlikely password that is memorable such as

Ilike88potatoesnot93carrots!!

change your password regularly, don't write your password down, and never tell anyone else what your password is.

As well as passwords, there are a number of hi-tech ways of identifying yourself to a computer. One area that has attracted a considerable amount of attention is biometric identification. Here, technologies that recognize a computer user from unique physical traits are employed, often in conjunction with passwords. Typical traits that can be used include fingerprints, iris patterns, a user's face, their voice, and the geometry of a hand. Many of the techniques are in the research phase with some having unacceptably high error rates and also being susceptible to attack. For example, an active area of voice technology is to simulate a person's voice from a set of recordings.

Non-technological security

In the advice I quoted from the Pfleeger and Pfleeger book *Security in Computing*, there was the injunction not to tell anyone your password. This is a good example of a precaution against non-technical security breaches and the problems that computer users face over and above those associated with viruses and other illegal technologies.

The former security hacker Kevin Mitnick, in his book *The Art of Deception*, describes a number of attacks on computer systems which require no technical knowledge at all. A number of hackers have used a variety of techniques categorized as social engineering to penetrate a computer system – not by employing worms, viruses or operating system trapdoors, but by taking advantage of human frailties. In his book, he states that social engineering

> ...uses influence and persuasion to deceive people by convincing them that the social engineer is someone he isn't, or by manipulation. As a result, the social engineer is able to take advantage of people to obtain information with or without the use of technology.

Here's a typical social engineering attack decribed by Mitnick. A potential intruder builds up a relationship with the clerk at a video rental store that forms part of a collection of video stores by masquerading as a clerk at another store. The relationship is nurtured for a number of months to the point where the innocent clerk assumes that the person they are talking to actually works at a sister store. Then, one day, the intruder rings up the clerk claiming that their computer is down and asks for details of a customer (name, address, credit card) who wants to borrow a video. Once obtained the customer's bank account can be rifled.

As well as crimes being committed using non-technical means, data have also been lost because of human frailty. The last decade has seen a large number of data leaks from companies and organizations. For example, in 2008, the British banking group HSBC admitted that it had lost data on many tens of thousands of its customers' insurance records. Although the disk was protected by a password, it was not encrypted, leaving it open to anyone with a good level of computing expertise to read it.

In 2007, in one of the largest data leaks in British computing history, the government's Revenue and Customs Department lost 25,000,000 records containing financial and other data belonging to British citizens who were being paid child benefit. An employee had written the details onto computer disks and had sent the disks unrecorded to another government department – they were then lost in the post.

There is an impression amongst the public that the main threats to security and to privacy arise from technological attack. However, the threat from more mundane sources is equally high. Data thefts, damage to software and hardware, and unauthorized access to computer systems can occur in a variety of non-technical ways: by someone finding computer printouts in a waste bin; by a window cleaner using a mobile phone camera to take a picture of a display containing sensitive information; by an office cleaner stealing documents from a desk; by a visitor to a company noting down a password written on a white board; by a disgruntled employee putting a hammer through the main server and the backup server of a company; or by someone dropping an unencrypted memory stick in the street.

To guard a computer against technical attacks is relatively easy: it involves purchasing security software and making sure that the files that the software uses – for example, virus signature files – are up to date. There is, of course, always a struggle between the intruder that uses technical means to access a computer and the staff who are responsible for security. Sometimes an attack succeeds: for example, denial of service attacks in the 1990s was a serious problem, but, very quickly, technical means have been developed to cope with them.

For companies and other organizations, it is much more difficult to guard against non-technical attacks: it requires a whole set of procedures which guard against all possible security risks. Examples of such procedures are those for handling visitors, such as enforcing the rule that they do not wander around a building by themselves; for disposing of waste computer printouts; for ensuring the safety of laptops when travelling; for disposing of old computers that may have sensitive data stored on their hard disks; for prohibiting the publishing of personal details on social web sites that may help social engineering attackers gain access to a computer network; and for ensuring a clean desk policy.

Non-technical security is much more difficult because it is much more pervasive than technical security: it requires the cooperation of everyone from reception area staff to the head of the computer centre, and, in organizations that take security seriously, is embedded in a thick document known as a security manual or manual of security procedures.

For the individual working at home who wants to guard against non-technical attacks, it is much easier. All that is needed is to encrypt any file that is sensitive, for example a spreadsheet containing financial data; employ adequate passwords for any files of sensitive data; never give data such as passwords and bank account numbers over the Internet or in response to a phone call; and, if you do have to carry data around, buy an encrypted memory stick – they used to be somewhat expensive, but have come down in price since they first came on the market.

Chapter 6
The disruptive computer

Borders UK

This morning, as I started writing the first draft of the first chapter of this book, I heard that Borders UK, one of my favourite bookshops, was in financial trouble – four days later, they went into administration. One reason quoted in the BBC news was that the company was finding it very difficult to compete with online retailers such as Amazon. It is much more expensive to run a chain of bookstores, some of which are situated in prime shopping locations, than a warehouse, a web site, and a call centre. Just as a comparison I looked at the price of a book that I bought recently. In a city-centre shop, I would pick it up at £34, while it was currently advertised on the Amazon web site at £20.13.

While the employees of Borders are able to use computers to help them in answering customer queries and ordering out-of-stock items, in the end they have threatened their jobs. I did think that bookstores had something of a future, since they offered browsing facilities, until I saw that a number of booksellers, via a system known as Google preview, allow visitors to their web site to browse the table of contents of a book and many of the pages within the book. *For the book trade, the computer is disruptive.*

My favourite record shop in Milton Keynes was the Virgin Megastore which, after a management buyout, renamed itself Zavvi. Shortly thereafter, the Zavvi retail chain folded. One of the reasons quoted was the availability of cheaper music downloads from sites such as iTunes. Again, the staff at Zavvi found the computer useful in carrying out their jobs, but, in the end, it killed off their trade. *For the music trade, the computer is disruptive.*

I can read the national newspapers from my computer; occasionally I do, but most of the time I access the BBC news site, a site that has won a number of awards for its design, implementation, and content. The increasing availability of news on the Internet has had a dramatic effect on newspaper sales. Newspapers across the Western world have been coping with a slide in advertising revenue, declining circulation, and a movement of readers to free news online. According to the USA Audit Bureau of Circulations, the average daily circulation for 379 daily American newspapers was down 10.62% in the April to September 2009 period, compared with the same period in 2008. The computer has helped the reporters who file copy for their newspapers: word processors are a fantastic piece of software. However, the computer has led to layoffs and staff reductions in many newspapers. *For the newspaper industry, the computer is disruptive.*

Another disruption associated with newspapers is the reduction in the amount of space devoted to reviews – for example, book reviews – and the fact that review staff, both permanent and freelance, are often the first to feel the chill winds of redundancy. Book reviewers and film reviewers used to be very powerful figures: their opinion could kill off a book or film or elevate it in the popular listings. There are now a large number of sites that review films and books. Some of them are just devoted to reviews – a site such as rottentomatoes.com – or provide reviews as part of their business: a good example here is the Amazon web site which, although it is devoted to the selling of a variety of goods, includes reviews of the goods in their catalogue – the reviews being written by customers.

In this section of the book, I shall look at some of the ways that computers have affected us in terms of how we interact with others, in terms of employment, and in terms of how technology is improving our lives; I will also look at some of the ways that might affect us negatively.

Disruptive technologies

One of the key writers about the disruptive effects of technology is Clayton Christensen. His two books, *The Innovator's Dilemma* and *The Innovator's Solution*, look at how technological devices such as the hard disk storage unit can have major effects on industry.

Disruptive innovations can be placed in two classes: low-end disruptive innovations and new-market disruptive innovations. The latter is where a technological advance creates new business opportunities, changes the behaviour of consumers, and often leads to the elimination of industrial sub-sectors. A lower-end disruptive innovation affects current technological objects and services and reduces their price and consequently their availability.

The computer has given rise to disruptions covered by both categories. For example, when the mobile phone was designed and developed the companies that manufactured them added messaging as an afterthought, not thinking that it would be a feature that would be used so much, to the point where it is inconceivable that a mobile telephone would now be sold without text messaging facilities. This is an example of new-market disruption which obliterated much of the pager industry.

An example of a lower-end disruption is that of open-source software which I discuss later in this chapter. This has enabled computer users to use operating systems, word processors, spreadsheet systems, and photo utilities which cost nothing – a true reduction in price.

Open-source development

When you use a computer to access the Internet, you have a possible audience of hundreds of millions of users and a potential to link those users together. Wikipedia is an example of mass collaboration: all the entries have been initiated and edited by volunteers. Another, equally impressive, example is the rise of open-source software. This is software that differs from commercial software in a number of ways; first, it is free; second, the program code of open-source software is available for anyone to read; and – this is where the term 'open source' comes from, 'source' referring to the program code – third, anyone can take the code, modify it, and even sell it as a product.

There is a large quantity of open-source program code in existence; the two most used are Apache and Linux. The former is the software that is used to dispense pages from a web server. Currently, it is estimated that something like 65% of web servers use Apache. Linux is an even more impressive story. It is an operating system, and a competitor to the popular Windows operating system. It has its roots in a simple operating system known as MINIX which was developed by a computer science academic, Andrew Tanenbaum, in order to teach his students about the design of large systems.

Linux was developed by Linus Torvalds, who was a student at the University of Helsinki. He decided to improve on MINIX and started developing Linux, initially inspired by some of the design ideas used for MINIX. The original implementation of Linux was so good that it came to the attention of a number of software developers at the same time that the Internet was blossoming. The result was that the system evolved into being maintained by programmers who devote their time for nothing.

Linux is very impressive in terms of both penetration and in terms of what it offers. A recent survey showed that 80% of the most reliable web-hosting companies used Linux and that it is the operating system of choice for many supercomputers. There is also a desktop version of Linux which contains the usual array of software that you would expect: word processors, graphics programs, image editors, email programs, and audio players.

In desktop terms, Linux still has some distance to go before it threatens the very popular Windows operating system. For example, it is sometimes fiddly to install new software for the desktop version. But it is still the most impressive example of the trend of computer users creating a large, complex artefact outside established institutions.

Advertising

The computer has had a major effect on the revenue earned by both television and newspaper companies. This is mainly down to the use of online adverts, but is also due to the fact that television programmes containing adverts can be recorded and, when wound back, the adverts can be fast-forwarded. The key to the disruption that has occurred in these industries is targeting. This concept is nothing new: all it means is that you place your adverts where you think they will get the maximum readership and payback. For example, if you had an advert for study tours to Greece which involved visiting ancient sites and listening to renowned Greek classicists, then you would probably have targeted the advert at a newspaper such as *The Independent*.

The computer and the Internet have changed the face of targeting and made it more effective. The major technology that has been disruptive is AdWords, a Google technology that earned that company billions of dollars. AdWords is based on what is known as a pay-per-click model. An advertiser who wants their products or services advertised on Google specifies an amount that they will pay when an online advert appears in a page retrieved when someone carries out a Google web search.

The advertiser specifies what keywords will trigger the adverts, so that if a user types the word 'Angling' into the Google search box, they will find, as I did when I typed in the word, that three adverts for online fishing tackle sites were displayed, one advert for fishing holidays displayed, an advert for angling items at Amazon shown, a link to a site that marketed a revolutionary way of throwing fish bait into a river or lake displayed, and a link to a site which collects together links associated with angling displayed.

AdWords represents a much more fine-grained approach to advertising than is found in conventional media. Its effect has been marked. In the third quarter of 2009, advertising revenues earned by American newspapers dropped by 28% as compared with the corresponding period in 2008; revenues from the first nine months of 2009 also dropped by 29%. Clearly, a component of this decline is the credit crunch problems experienced by American firms in these years; however, it is an important part of a trend that has occurred since the early part of the decade.

The television industry has also suffered drops in income. For example, according to analysis carried out by *The Times*, Google earned £327 million in the United Kingdom compared with £317 million for all of the British commercial channel ITV1's output during the period between July and September 2007.

IT outsourcing

In 2011, my BT Internet connection failed; this was, for me, the equivalent of having my electricity or water cut off (something discussed in the final chapter of this book when I look at the work of Nicholas Carr). I rang the BT helpline and an Indian voice replied. Over the next 20 minutes, using the Internet, he took control of my computer. He opened folders, changed some settings, and restarted some programs, and, lo and behold, my connection restarted. I'm not often agog, but this time I was. I watched my mouse pointer travel over my screen guided by an invisible hand – I had the feeling I get when I drive past Heathrow airport and see a Boeing 747 taking off: a feeling of knowing that what I saw was possible, but not quite believing it. For the record, the fault was not BT's: I was in the habit of switching my Internet modem on and off and, during the off period, it had missed an important update.

This is an example of outsourcing. In a number of conventional industries, outsourcing has been the norm. For example, much of the clothing sold in the United Kingdom has been manufactured in countries such as China, India, and the Dominican Republic; and electronic devices are often made in low-labour-cost economies, for example my iPod is made in China. However, outsourcing is now common in systems development.

Pinsent Masons, one of the United Kingdom's leading law firms, have listed the arguments for using an external software developer: lower costs because of efficiencies and economies of scale; access to high-level IT skills (for example, software developers in India are some of the most accurate and sophisticated coders in the world and use advanced tools for systems development; they also have the largest proportion of companies certified to produce software of the highest reliability); removing non-core business from a company's infrastructure; minimizing large capital expenditure on IT infrastructure; and having some degree of certainty of future IT spend.

The Internet has provided an infrastructure that enables customers to talk to systems analysts in other countries via video links, send documents via emails, and test systems via live links. The result is that computer-based development is increasingly being moved offshore. For example, in 2006, the research company Computer Economics reported that 61% of all the American companies they surveyed outsourced some or all of their software development.

As I sit typing this chapter, three years after one of the greatest upheavals to our financial system, the future of IT outsourcing is unclear; on the one hand, companies are cutting back on IT investment, on the other hand, offshore outsourcing companies offer significant savings in investment. These two factors, together with increases in IT skills levels, will determine the growth of outsourcing over the next few decades.

The type of outsourcing I discuss above is thick-grained outsourcing in that you hire a company to carry out some set of IT functions. There is also a much thinner-grained version. There are now a number of web sites that provide details of software staff for hire (often these programmers are from Eastern Europe). So, if you have a project that requires a small amount of software development over a limited time, then video links and emails can put you in touch with developers who are competitive compared with Western European rates.

Citizen journalism

The term 'citizen journalism' is used to describe how ordinary citizens have become empowered by cheap computers and Internet connections to involve themselves in the reporting of reaction to events and publishing news articles and opinion articles. There are a number of technologies that are used for this: the main ones are blogs (online diaries), podcasts, video footage, digital photographs, and web sites.

There are a number of manifestations of this phenomenon. There are news blogs which aggregate news and contain comments not just by the blogger, but also by other Internet users who are able to insert comments at the end of the blog text. These blogs can be general in content or specific to a particular topic such as technology.

There are also news web sites which contain the same type of material that would be found in the web site of a conventional newspaper. Some of these sites take a neutral viewpoint; often, though, they report and comment on the news from a particular position, for example from the point of view of the Green movement.

Some of the most interesting manifestations of citizen journalism are participatory news sites where computer users post articles, other users tag the articles with descriptors that provide easy indexing, and, on some sites such as Digg, users vote on which articles are interesting. The articles with the greatest number of posts are then promoted to a prominent position on the web site.

One of the results of the drop in price of computers and silicon circuits has been the accompanying drop in price and availability of digital devices. One of the areas where this has been most marked is in digital recording. To my left, while I write this chapter, I have a digital recorder made by the Marantz company. It costs about £400, produces sound quality comparable to the recorders used by radio interviewers, and is a fraction of the cost and more convenient than the tape-based recorders of the preceding decade. Anyone can buy two condenser microphones, a cheap sound mixer, and one of these recorders for about £800, and can turn themselves into an Internet radio station. Many do.

The computer and the Internet have provided a medium for free expression that was only available to journalists up until the beginning of this decade. The most that anyone could have expected previously was a letter to the editor which might or might not have been selected. Now, as soon as there is a major story, the Internet is swamped by traffic.

An example of this effect happened in November 2009. A hacker successfully accessed a computer used by the Climatic Research Unit at the University of East Anglia. This is one of the foremost global warming research units in the world. Within days, there was a maelstrom of emails, audio podcasts, video podcasts, blog entries, and news stories circulating within the Internet. The director of the unit stepped down while an independent review looked into the claims.

Five days after it appeared, I did a search on Google using the word 'Climategate', the term used by climate sceptics to label the incident. There were over 13 million hits. I also checked on the video YouTube site, and there were over 16,000 hits when I typed in the same keyword.

It is clear that in one direct way (its ability to spread news) and one indirect way (its lack of security) the computer has had and will have a major effect on journalism.

Digital photography

For a long time I used film cameras. I would buy a film – usually containing enough space for 36 exposures – at the local chemist or photographer's, and load it into the camera, take the photographs, and then return the exposed film to the chemist for developing. The chemist would then send the film to a developing laboratory and, eventually, I would collect my 'snaps'. Hobbyists would sometimes develop their own photographs; for this, they needed to soak the film in a variety of chemicals and then take the negatives and place them in a device known as an enlarger. This would then produce an image on a sheet of paper soaked with chemicals. The paper would then be developed in a number of trays containing more chemicals. All this took place almost in the dark using just red light.

I now take photographs with a digital camera, and at the setting I use, I can take hundreds of photographs. All I need do to look at them is to load them into a computer and use one of a number of image viewer and manipulation programs that are available.

A hobbyist who wanted to modify a film-based image would have to carry out some complex, error-prone manipulations; for example, they would have to pass cardboard shapes between the beam of the enlarger and the photo in order to restrict the light and hence alter the light values of the image. Modern photograph-manipulation programs such as Photoshop provide facilities that enable the photographer to manipulate a photograph in a variety of ways, for example changing colours, changing exposure on the whole of a photograph or part of a photograph, tinting, creating effects similar to those achieved by post-Impressionist painters, and picking on areas in a photograph and enlarging them.

In 2002, the sales of digital cameras surpassed those of film cameras. Whenever I go to a tourist spot I now very rarely see a film camera in use. The trend since then has been for digital cameras to have better and better facilities, such as a greater resolution. This change is due to hardware improvements, faster processors, the increase in density of pixels in image sensors, and a reduction in circuit size. There has also been an increase in the facilities photographic software offers. For example, there is a technique known as High Dynamic Range imaging (HDR) that can be used to produce ultra-realistic photographs which are combined from a number of versions of a digital photo that are taken with different exposure values.

There has also been a trend for digital cameras to be embedded in other devices, primarily mobile phones. The growth of mobile phones has been staggering. In 2002, there were approximately 1 billion mobile subscriptions and one billion fixed-line subscriptions in the world. In 2008, there were approximately 1.27 billion fixed-line subscriptions and 4 billion mobile subscriptions. The vast majority of these phones in the Western world feature a camera. One effect of this is that it has created a citizen journalist movement. For example, Janis Krums produced the first photo of US Airways flight 1549 in the Hudson river after ditching because of engine problems. He took the picture on his iPhone minutes after the plane ditched in the water – a dramatic photograph showing passengers standing on one of the plane's wings while others huddled on the floating emergency chute. In June 2009, there were major protests in Iran about the alleged rigging of the presidential election; many of these protests were broken up violently by the authorities. Pictures showing this could be found on the Internet minutes later, as protestors sent photographs taken with their phones to friends outside Iran.

There are a number of other effects of the rise of digital photography. One obvious one is the demise of the chemical developing laboratories. The increasing power of computers and their increasing miniaturization has meant that such laboratories had no easy upgrade path to change themselves into digital laboratories, since the local photography shop or chain chemist can now invest in sophisticated digital photographic equipment – all that is left now are a few very specialist firms which cater for the fine art industry, for example photographers who take highly detailed black and white photographs.

There are some less obvious effects. A number of cultural commentators have raised the problem of trust in photographs. For example, political opponents circulated a digital photo of presidential candidate Senator John Kerry next to Jane Fonda at an anti-Vietnam war rally – the photo was doctored. There are a number of examples of institutions inserting faces of black people into photographs to give an impression that they are more raciably diverse.

All of this raises questions about which images we can trust. Clearly, this sort of manipulation could have been done with film technology, but it would be much less easy to do. To doctor a digital image using a computer program is far easier: it took me 4 minutes to transpose a picture taken of my wife in our village and add a background taken in Avignon to give the impression that she accompanied me on my trip.

Digital photography also has another citizen journalism effect over and above the use of the mobile phone. The Abu Ghraib incident arose from the sending of digital photographs of American soldiers mistreating Iraqi prisoners to friends and acquaintances.

The digital camera also provides more opportunities for creativity. Artists such as John Simon, Shawn Brixey, and Pascal Dombis routinely use the computer in their work. Dirck Halstead, a professor at the University of Texas, carried out a survey of the readers of the magazine *The Digital Journalist* that asked the readers about their attitudes to digital camera technology. All of them preferred digital technology; one surprising result from the survey was that the creativity that it offered was more of an advantage than factors such as speed and convenience.

Science and the computer

One of the startling things happening in science research is that it is being deluged with data. Computers are now routinely being used to store and analyse billions of items of data from satellites orbiting the earth, from land-based weather stations, and from large-scale nuclear experiments.

As an example of this, the Large Synoptic Survey Telescope, which is due to come into operation in the next ten years; it uses very sophisticated computer hardware, hardware similar to that used in digital photography, and will, in its first year of operation, produce something like 1.3 petabytes of data, more than any other telescope has provided by a very large margin.

This flood of data is providing scientists with major opportunities to analyse and predict, for example predicting the effect of high rainfalls on flooding near major centres of population. However, it is also providing some major problems. There are large numbers of diverse research groups throughout the world producing data. If a research group, say in atmospheric physics, wants the data and program code from another group, the normal process is to send an email, and if another group wants the same data, then the same process occurs. So the first problem that scientists are facing is that of distributed data.

The second problem is that there is no acceptable standard for storage. There are two components associated with the storage of data. First there is the data itself: large amounts of floating point numbers separated by symbols such as spaces or commas. Second, there is metadata. This is data that describes the raw data. For example, if the data were from a series of temperature measuring stations throughout the globe the metadata would identify the part of the raw data that specified the station, the periods over which the data was sampled, and the accuracy of the data. Currently we have no standard way of specifying these two components.

The third problem is that it is not just data that are important in scientific enquiry. For example, when a physicist looks at data that are being used to prove or disprove global warming, they are analysed using computer programs, and embedded within these programs is code that carries out some statistical processes. It is very difficult to develop completely error-free computer programs. If other scientists want to audit their colleagues' work or progress it, then they will need access to program code.

The advent of the embedded computer – embedded in measuring instruments and observational instruments such as radio telescopes – has been disruptive to science. Clearly there is a need for much more attention to the archiving of data. There are some initiatives that address this such as the Australian National Data Service, but much more is needed.

The long tail

In 2007, Chris Anderson published the book *The Long Tail*. Anderson, a staffer at *Wired* magazine, pointed out that if you

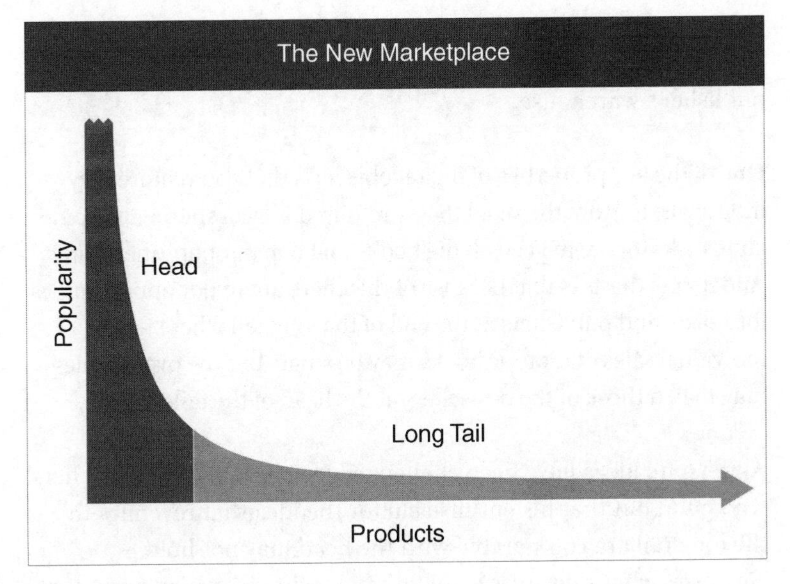

7. The long tail

looked at the sales of digital objects such as books you would see the graph shown as Figure 7. The horizontal axis shows books while the vertical axis shows sales. What you see is a standard curve for items that have a number of different instantiations, for example books, DVDs, and computer programs.

What the graph shows is that there is a comparatively small number of books that have large sales at the head of the graph and that there is a long tail of books that have small sales.

The key idea behind Anderson's thesis is that digitization promises the seller of digital items as much profit in selling the items in the tail as there is in selling items at the head. In the past, booksellers were hampered in what they sold by expensive stocking costs and the physical limitations of their shops. Publishers had similar constraints: every few years, I would receive a polite letter from one of my publishers informing me that one of my books is going out of print and would I like to buy some of the remaining stock.

Going out of print effectively means there will be no more copies available via conventional bookshops and nothing in the publisher's warehouse.

One of the key properties of digital objects is that you require very little space to store them: all that is required is a computer and some large disk storage to keep all the books that a major publisher prints. Anderson's thesis is that because of this there are major opportunities for sellers and publishers at the end of the long tail where the individual sales of esoteric books may be small, but the overall sales may match those of the bestsellers at the head of the tail.

Anderson's ideas have been challenged by a number of researchers who point out that his enthusiasm for the idea that revenues in the long tail are comparable with the head may not hold. However, what they miss is that digital technology may mean that esoteric titles such as *Steam Trains of Romania in the Early Part of the Twentieth Century*, or a DVD on early Japanese horror movies that feature zombies, might not match the sales of the latest bodice ripper, but they do offer, along with many other esoterica, opportunities for increased sales. For the major publisher and the book chain, the computer is disruptive in a positive way: for the specialist bookseller sourcing rare, non-antique works, the disruption is in the other direction.

Books and the e-reader

When I go for my lunch, I pass a colleague who has a small, flat, computer-based device in his hand. It's an e-reader. An e-reader allows him to download books from either a publisher's web site or a bookseller's site. I saw a prototype e-reader in 2005 and was not impressed, mainly because the quality of the screen was poor. My colleague's e-reader is vastly superior. The emergence of the e-reader reflects a trend that has happened in music whereby vinyl records gave way to CDs and where CDs are steadily giving way to sound files such as those that are played by the iPod. There is an interesting debate going on about the future of e-books and e-readers, the result of which will determine whether, in a decade or two, we will see the end of the conventional bookshop.

On the one hand there are those who point out that many people buy books for a quasi-aesthetic experience, that while reading is obviously the main purpose in buying a book, a well-produced art book, for example, with glossy pages and high definition, is an artefact that is pleasurable to handle and to look at. There are also those who point out that the price differential between online books and their printed version is not that large; this is due to the fact that most of the cost of a book comes from its marketing, its editing, the profit that a bookshop makes, and the profit that the publisher makes, and because of this small price differential there will be no major shift from printed media into electronic media.

However, there are those who compare the growth of e-books to the growth of music files who state that the same differentials hold between online music and CDs. They also point out that text is a static medium and that e-books that feature video and audio clips and links to web sites would provide a much richer reading experience – one that would drive readers towards e-books.

It is early days yet – however, the indications are that the growth of e-books is accelerating. A survey by the Association of American Publishers in conjunction with the International Digital Publishing Forum recently showed an exponential growth. For example, wholesale revenue from e-books in the third quarter of 2008 was 14 million dollars, while in the corresponding quarter of 2009, it was around 47 million dollars. I do not want to make a case that these figures show that in *n* years e-books will dominate; after all, the sales of conventional books greatly outweigh the sales of e-books. For example, the Association of American Publishers reported sales of 1.26 billion dollars for conventional books. However, the data shows the sort of trend that occurred with the increasing availability of sound files for devices such as the iPod.

In late 2009, *The Bookseller*, the trade magazine for British bookshops and publishers, released the results of a survey into the commercial possibilities of e-books. They were quite surprising: although 88% of the thousand-odd respondents stated that they felt that the e-reader was a threat to their business, many felt that there were commercial opportunities. One typical response was

> Everyone will gain by making reading easier and more accessible – and by widening the appeal to younger people (i.e. mobile audiences). High street book shops need to become service providers for readers – technology, some printed books (e.g. children's books, maps, art books), advice, author readings, seminars, learning centres, event hosts, etc.

One scenario is that the conventional chain bookshop disappears under pressure from online competition, but the local bookshop makes a comeback by becoming a venue and social centre and perhaps a micro-publisher.

My view is that it is not clear what will happen with respect to the balance between conventional book sales and the sales of e-books. However, here is one scenario. That there will be an increase in e-books that lie outside the considerable categories of fiction and non-fiction that have a personal viewpoint. For example, a prime area for e-books is encyclopaedias and dictionaries. The online project Wikipedia has already shown the way here: each entry not only contains text on a Wikipedia article, but also cross references other articles – often too much – and contains many cross-links to relevant items such as papers, news articles, and blogs.

Another area where e-books could take off is travel books where, as well as conventional text, the book would contain items such as video clips and photographs. As an experiment, I typed in the keywords 'Musee d'Orsay', a fantastic art gallery in Paris that has been converted from a railway station that holds some of the greatest 19th-century and 20th-century paintings. I got 47,000 hits. I looked at the first two hundred and, with a few exceptions, the quality of the photography is excellent and provides a good indication of the variety and importance of the holdings in one of my favourite art galleries. I also checked on the video hosting site YouTube and received 1,800 video clips.

Another area that could experience explosive growth is that of instruction manuals, for example cookbooks and car-maintenance manuals. Here, video clips showing some of the more tricky cooking techniques would be interspersed with conventional text and photographs.

There are many other categories of book that could be transformed by e-book technology. For example, school textbooks could contain video podcasts explaining a difficult point. However, at the end of the day, there is a huge edifice that e-books need to address: that of the novel and the non-fiction book which has a personal viewpoint, such as a single-writer history of the Second World War.

There are a number of factors that may accelerate the trend. The first is the availability of free material – open source text if you like. Project Gutenberg is a project that has a long history: it started in the early 1970s when its founder, Michael Hart, typed the Declaration of Independence into the mainframe computer at the University of Illinois. Since then volunteers have typed and digitized over 30,000 books into the Project Gutenberg web site. All the books are out of copyright and contain a high proportion of esoterica. However, there are many novels written by authors such as Dickens, Forster, and Scott Fitgerald which can be loaded into an e-reader – legally and for free.

A wild card that may affect the growth of novels and non-fiction work is Google. Over the last six years the company has been digitizing books on an industrial scale. This has been done in conjunction with libraries around the world such as those at

Columbia University and Cornell University. In 2009, the current count of Google digitized books stood at 10 million. Most of these books were available for partial search but around a million could be read in their entirety. The spread of Google books, the alacrity with which publishers now produce electronic versions of paper books and the technological advances that have produced a new generation of e-readers has created a major threat to paper books.

Another factor that may drive the cost of books down is that of online advertising. Every year I go to the Oxford Literary Festival – usually with my wife. Last year we attended a session in which two of my favourite authors were interviewed and fielded questions. They were Donna Leone, who writes crime novels based in Venice, and Kate Atkinson, who writes high-quality, uncategorizable novels. Before we attended their session, I predicted to my wife that the vast majority of the audience would be women and middle class. A quick head-count confirmed that about 95% were certainly women – I suspect that the same proportion were middle class. This congregation is a marketer's dream.

One of the key tools of marketers is targeting: knowing what the demographics are of the potential audience for some media item: a TV programme, a newspaper, or a TV channel. Most targeting is broad-based: the audience for Channel 4 in the United Kingdom has a number of broad traits as does the readership of *The Daily Telegraph* and you find advertising addressing these.

The purchasers of a book are a much better-focused target than, say, the readers of a particular newspaper. The purchaser of a book on the natural world may be interested in other books that are similar to the one bought (a trick that Amazon uses to display possible books to a visitor to their site), they may be interested in holidays which involve activities such as observing wild animals, or they may just need a new pair of binoculars.

The e-reader sold by Amazon already has Internet access built in, as has the iPad – a multi-function device developed by Apple. Soon all e-readers will have this capability. It raises the prospect that e-book prices could be driven downward for those who want to purchase a title that has advertising on part of the screen, where the cost of the books is reduced by advertising revenues.

Chapter 7
The cloud computer

Introduction

Whenever my wife or I visit out local Tesco supermarket we take part in an implicit contract: we buy our groceries and, before handing over our credit card, we give the checkout person our Tesco Clubcard. This card is swiped and the computer in the checkout till sends the card number and the items that we buy to a central computer. The use of such cards has revolutionized marketing. Previous to their use the information from a checkout computer was aggregated information, for example, how much of a certain product was bought in a week and whether there was an increased demand for a certain product at specific times of the year. By being able to associate a collection of purchases with a particular customer or family group of customers, a supermarket is able to carry out some very canny marketing.

Tesco will know that: my family buys lots of fresh food, that if it is a Wednesday we will buy some pre-cooked meals, and that red wine is the *drink du jour* of our house. I am relaxed about the data that is held: I gain an advantage because it enables Tesco to send me special offer vouchers for items that they think I would like and buy more frequently; it also adds to the Air Mile travel points that come with the Tesco Club Card; and Tesco get an advantage in knowing what my family spending habits are. It is a deal that I am not prepared to make with a pharmaceutical retailer since purchases there are associated with health.

Loyalty cards were the first manifestation of an explosion of data; an explosion that the Internet caused and which can be used by both companies and individuals with access to a browser. The aim of this chapter is to look at how the Internet has transformed the computer to the point where when you refer to the web as a massive computer you cannot be accused of science-fiction foolishness.

Open and closed data

The Tesco data are an example of closed data: the only people who can see it are marketers employed by that company. There is plenty of closed data that has been generated by users of the Internet. A good example is the data collected by the Amazon web site. Whenever I log in to that site I am greeted by a front page that displays books that Amazon thinks that I like. The books that are displayed are based on my past buying habits. For example, if I bought a book on contract bridge one week, then the web site will point me at other books on this card game.

Such closed data is valuable to large companies that make their money from selling or hiring items to the general public.
A startling example of the lengths companies will go to improve the processing of this data is the Netflix Prize. Netflix is a company that rents out and streams videos to customers. It has a large database of customer preferences and tastes, and it was able to predict that if a customer selected a particular DVD, then there would be a good chance that they would want to borrow similar DVDs. The company used a computer program to make the predictions; however, it was felt that the program could be improved, and what Netflix did was to set a challenge to the programming community to come up with a better program. The prize for this was $1 million. In 2009, the prize was won by a team

of programmers, the majority of whom worked for American and European systems companies.

As well as closed data, there is a huge quantity of open data that can be accessed by anyone who can use a browser. In 2000, the Canadian gold-mining firm Goldcorp placed its crown jewels on the Internet for everyone to see.

The company had hit bad times: its current holdings were looking mined out, the price of gold was falling, and the company lurched from crisis to crisis. Goldcorp's chief executive officer Rob McEwen decided to make the company's exploration data – its crown jewels – public and run a competition in which members of the public would process the data and suggest future exploration sites. Goldcorp offered $575,000 in prizes to the best entries.

It is not clear what the motives for McEwen's radical move were: desperation or an insight into how the power of crowds could be harnessed. Whatever the reason the competition was a huge success. A wide variety of people entered it: military men, academics, working geologists, retired geologists, and hobby geologists. The result was that new, very productive sites were found. The competition transformed Goldcorp into a major player in the gold exploration market and increased its value from $100 million to many billions of dollars.

The Goldcorp competition marked the start of an explosion in the availability of open data to the public; not only data that could be gained from processing public web pages, but data that would normally be hidden away in the computer networks of a commercial concern.

An example of public data is the online encyclopaedia Wikipedia. This has become one of the major successes of the Internet; it contains over 2.5 million entries all contributed by Internet users and is an example of the collaborative spirit that I will describe in this chapter. What is not generally known about Wikipedia is that it is a major resource for researchers and commercial developers who are concerned with natural language processing, for example companies producing programs that summarize thousands of words into an easily digestible form and scan the Internet for texts – newspaper articles, reports, magazine articles, and research papers – that deal with a particular topic.

One of the problems with natural language texts is that of ambiguity. The classical example of this, quoted in many textbooks on the subject, is the sentence

> They are flying planes.

This could have a number of meanings. It could refer to the pilots of commercial planes or pilots of planes flown by members of the armed forces. It could also be used by someone pointing at a number of planes in the sky. It could even be used to describe children flying model planes.

By itself, this sentence is difficult to understand. However, what makes it understandable is its context. For example, if it occurs in an RAF recruitment brochure, then it would have the first meaning. Researchers have found that Wikipedia provides a very useful resource for tasks such as understanding text. In order to understand this, look at the extract below from the encyclopaedia.

> Public cloud or external cloud describes cloud computing in the traditional mainstream sense, whereby resources are dynamically provisioned on a fine-grained, self-service basis over the Internet, via web applications/web services, from an off-site third-party provider who shares resources and bills on a fine-grained utility computing basis.

The extract comes from an entry on cloud computing which underlies much of this chapter. The underlined terms within the extract are references to other Wikipedia entries. It is these which provide a context for understanding a sentence. For example, if a sentence in an article contains the word 'cloud' and many of the cross references in the Wikipedia article on cloud computing occur in the article you would be able to identify it as one about a particular brand of computing rather than meteorology.

Another example of open data is that generated by Twitter. This is a technology that allows its users to deposit short messages on a web site that others can read; it's a bit like the messaging you get with mobile phones, the only difference being the fact that the messages (tweets) are public.

Tweets are being used by the US Geological Survey (USGS) to get public reaction to earthquakes. They provide a near instant feedback as to the severity of a tremor and enable emergency services to be organized a little more quickly than they would be via conventional monitoring.

The USGS continuously collects geo-codes (these are identification codes supplied by mobile devices such as 3-G mobile phones) and stores the tweets associated with the codes. When the US national seismic network detects an earthquake, a system then checks to see if there was a significant increase in messages following the event and the nature of the messages. Staff can then examine the messages to see what the effect of the earthquake was.

APIs

So, I have given you examples of databases that can be manipulated by a computer for a number of purposes. The important question is: how is this data made available? The key is something known as an Application Programming Interface (normally abbreviated to API).

An API is a collection of software facilities that allow you to access Internet-based resources. For instance, Amazon has an API that allows programmers to query data about the products that Amazon sells. For example, you can retrieve the unique ISBN of a book, and other product data that can be used for commercial activities. This might seem an innocuous thing to do but, later in the chapter, I will describe how revolutionary this is.

There is a web site known as programmableweb.com which contains details of all the available APIs that can be used to access web resources. At the time of writing this chapter, there were just over 1,900 APIs listed.

So, what can you do with such facilities? A good example is that of someone setting themselves up as a specialist bookseller, for example selling books on France. Amazon has an associate program that provides facilities for anyone to advertise books stocked by the company on their own web site. Such a web site would contain links to the book entries on the Amazon web site and when a visitor to, say, the France book site clicks one of these links and buys the book Amazon pays the owner of the site a commission.

A good quality site with, for example, articles on France, recipes for French food, and latest news items, would draw visitors who would feel inclined to support the site by buying their books using the site links – after all, the price of the book would be the same as that displayed on the Amazon web site. The details on the web site about each book would be obtained using an API that is freely provided by Amazon.

This form of associate program – sometimes known as an affiliate program – is now very popular. The web site associateprograms.com lists many thousands of such programmes. It includes organizations in the media sector, the food sector, and the sport and recreation sector.

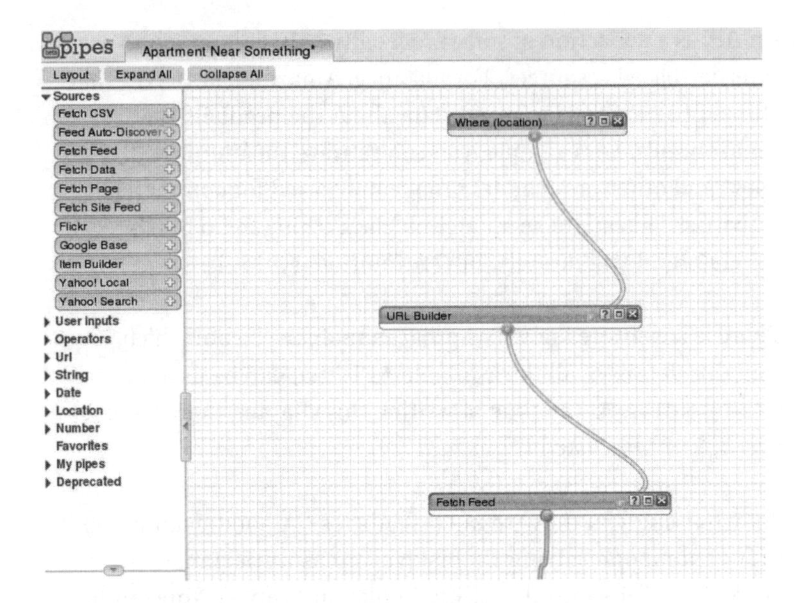

8. A Yahoo Pipes program

Another use of APIs is via something known as a mashup. The term 'mashup' comes from brewing where it is used to describe the bringing together of the ingredients for an alcoholic drink – usually beer. In computer terms, a mashup is the use of a number of APIs to create a hybrid application.

One of the first mashups was developed by the software developer Paul Rademacher. Google has stored a series of maps which can be viewed by anyone and for which it developed an API. Rademacher developed a program that took housing ads on the popular web site Craigslist and then displayed the location of the apartments or houses for sale on the maps published by Google.

There are many other examples. The web site programmableweb.com lists many thousands of examples with over half involving mapping. For example, it lists: a mashup which provides tourist guides for ten large American cities linked to videos on the YouTube site showing landmarks and visitor attractions, a mashup which allows visitors to share their most cherished books, and mashups of the location of crimes committed in American cities linked to the geographical location where they occurred.

Developing a mashup used to be something that was confined to software developers. However, the company Yahoo has developed a technology that enables anyone to develop mashups. The technology, known as Yahoo Pipes, is based on a very simple graphical programming language. It manipulates data feeds from web sites in order to create new applications.

Figure 8 shows an example of part of a Pipes program. To develop a Pipes program, all you have to do is to connect together a variety of boxes that carry out functions such as filtering data, recognizing data as important, passing it on to other boxes, and displaying data on a web page. While developing Pipes programs is not trivial, it is very much easier than many mashup techniques, and has lowered the technical barriers for anyone who wants to develop web-based applications.

Computers in the cloud

A number of the trends described in this chapter and in Chapter 1 point towards a vision of computing in the next decade as being highly centralized, with many of the functions carried out by computers such as the humble PC being transferred to a massive collection of server computers that reside in a virtual cloud. The last two years has seen an increase in facilities that reside in the cloud.

In the 1990s, if you wanted to use a software package, say an HR package which managed the staff in your company, then you would buy such a package and install it on the computers owned by your company. The package would normally be sent to you on some magnetic medium such as a CD or a DVD and technical staff in your company would then install it on any computers that needed it. Technical staff would then look after the software: installing new versions, tuning it for optimum performance, and answering any technical queries from users.

The increased speed of Internet access has meant that a different business model is emerging. Here the users of a package would not access it on their own computers, but use an Internet connection to access the software, with the software residing on a computer operated by the package developer. There would be a number of advantages in doing this.

The first advantage is that it would require less technical staff, or at least enable staff to concentrate on other tasks rather than spending time being trained and keeping up-to-date on the arcane knowledge required to maintain the package.

The second advantage is that it reduces the effort required to maintain versions of the software. Large software systems are often updated, for example when enough customers require new functions or when a law is changed that requires a company to behave differently. The process of installing an update can be something of a technical nightmare for staff who don't have the technical skills. By having only one copy of the software resident at a single computer, the update process is much more straightforward.

A third advantage is that computer hardware costs are reduced. The cloud computing concept envisages both the software and the data it requires reside in the cloud. Theoretically all that would be needed for a company to access a package in the cloud would be a very cut-down computer with little, if any, file storage – effectively what was known as a 'dumb terminal' in the 1970s.

The proponents of cloud computing claim that: it enables companies to be a lot more agile in their business, reduces both human and hardware costs, enables users to connect anywhere to software rather than just from a computer that resides in their office environment, enables a much higher hardware efficiency as a server on which a package resides can use up the slack resources I detailed in Chapter 1 more effectively, enables a more efficient use of resources, and addresses the green agenda. It also enables a higher degree of security because all the data that are used are held on a limited number of servers rather than being scattered around a large number of individual personal computers which are prone to both technical security attacks and attacks which arise from physical security problems.

There are also potential problems with cloud computing. First, what happens when a company becomes bankrupt? If no company is keen to take over the bankrupt company, it would severely embarrass the customers of the defunct company. Second, although there is an argument that security is enhanced when cloud computing is employed, there is the problem that if a security violation occurred at the cloud company's installation, then the effect would be much larger. If a single company had its data stolen, then that's a major problem for the company: if the data were stolen from a cloud company, *all* its customers would be compromised.

Cloud computing is not just confined to industrial and commercial packages, it is beginning to affect the home computer user. A good example of this is email. Many computer users have switched to smartphone-based email programs and to web-based mail systems. One example of this is Google Apps. This is a set of office-based tools which include a word processor, an email program, and a calendar. If you are happy with basic features, then these applications are free. Not only are the applications free, but also the storage that you use is free, for example the storage for word-processed documents.

One of the most interesting writers on the evolution of computing is Nicholas Carr. He has written two books which have divided the

IT community. The first is *Does IT Matter*. Here he argued that the strategic importance of information technology in business has diminished as information technology has become more commonplace, standardized, and cheaper. In this and a number of more academic articles, he pointed out in 2004 that as hardware costs decline, companies would purchase their computing almost as if it was a utility such as electricity and gas. Cloud computing is an example of this.

His second book, *The Big Switch: Rewiring the World, from 'Edison' to 'Google'*, takes the theme of utility computing further and uses the analogy of the growth of the electricity industry to shine a light on what is starting to happen in computing. Carr points out that the electricity generation industry started out almost as a cottage industry. A company that needed electrical power would install its own generator and have it maintained by its engineering staff; something which gave rise to a major demand for engineers with heavy electrical engineering skills. He shows that pioneers such as Edison changed the world of electrical generation to the point where electricity was generated and distributed by major utility companies.

In a series of elegant analogies, Carr shows the same forces at work in computing: where gradually services are migrating to large collections of servers that reside on the Internet rather than in some company premises. The implications of this, which I shall explore in the final chapter, include: a shift of control from media and other institutions to individuals, major worries about security and privacy, and the export of the jobs of highly skilled knowledge workers.

The social computer

In 1995, a web site was placed on the Internet. This site, known as WikiWikiWeb, has transformed the way we use computers. It was developed in order to help programmers in software engineering communicate with each other – effectively, it was the electronic version of the white board. Engineers could add ideas to the white board, delete anything they thought was irrelevant, and modify the thoughts of others.

Up until this site went live, traffic from computers to the Internet was mainly one way. If you wanted to view a web page, then you would click on a link, a short message of a few bytes would be sent to a web server, and a page containing thousands of bytes would be returned and displayed on your browser. WikiWikiWeb started a process whereby the interaction between a computer user and the Internet became much more two-way.

There were some early examples of two-way communication. For example, the online bookseller Amazon invited users of its web site to post comments on the books that they read. These comments were then inserted into the web page that contained details of the book. However, this interaction was not instantaneous: it was intercepted by a human editor for checking and then inserted.

The inventor of WikiWikiWeb was Ward Cunningham, a researcher and consultant. He wanted to be able to communicate with fellow workers in such a way that they could use a virtual white board. Cunningham's idea has spawned a software system known as a 'wiki'. What a wiki does is to maintain a document or series of documents within the Internet. Users of the wiki are then allowed to modify its contents – even being allowed to delete some content. Most inventions associated with the computer can be seen as being prompted by some improvement in software or hardware. The invention of the wiki was unusual as it was conceptually simple and required very little change to existing systems.

Arguably the best example of a wiki is Wikipedia. This online encyclopaedia has been developed by thousands of volunteers;

a controversial study carried out by the prestigious science journal *Nature* compared it with the equally prestigious *Encyclopædia Britannica* and came to the conclusion that they were close in terms of accuracy.

There are many applications of wikis. For example, companies use them for interacting with customers: eliciting views about current products and services and discovering what customers want. Another example is the use of a wiki for running large projects – particularly those that are split across a number of geographic locations. My own university – a university where the vast majority of students study at home – uses wikis to enable students to collaborate with each other; it provides a virtual form of interaction similar to the social interaction that would occur in a conventional university.

Wikis heralded an era of two-way communication that is now the norm on the Internet. The social networking site Facebook is a good example of this. Facebook is a site that allows anyone to join, set up their own personal pages, and communicate with groups of other Internet users who are known as 'friends'. One of the many facilities offered by Facebook is 'the wall'. This is a shared area where a user and their friends can write messages that they can all access and read.

Wikis are just one component in the increasing use of the computer for social communication and in the development of collaborative products. Twitter is a web site that allows very short messages (known as 'tweets') to be sent from anyone with a connection to the Internet – mobile or otherwise. The web site Flickr allows users to store photographs on the web in such a way that anyone with a computer and a connection can look at and comment on them. YouTube is a video-sharing site that allows users to post short videos on the Internet and, like Flickr, allows anyone to comment on them; delicious is a web site that allows computer users to share the bookmarks of favourite web sites. Digg is a web site that enables users to share links and stories from anywhere in the world. Users of Digg can then vote on the stories and they will be elevated to a more prominent position on the site (a process known as 'digging') or reduced in prominence (a process known as 'burying').

The collaborative development of products is also an area that has blossomed over the last ten years. Wikipedia is, yet again, a good example of this: the development of a hugely comprehensive encyclopaedia by a large number of volunteers. There are, however, a number of other examples.

One academic example is OpenWetWare. This is an MIT project that aims to share knowledge in biology and which, via a number of collaborating institutions, stores research outputs such as articles and experimental protocols.

The computer as a data-mining tool

One of the consequences of the connections between the computer and the Internet and advances in storage technology is the large amount of data that is generated, stored, and made readily accessible to computer users. In general, the larger the amount of data that is stored the more information you can get out of it – but often at a higher processing cost.

A recent example of data-mining is described in Steven Levitt and Stephen Dubner's recent book *Superfreakonomics*. They describe how the financial records and habits of potential terrorists can be mined for information; this includes negative as well as positive features, an example of a negative feature being they were unlikely to have a savings account and a positive feature being the fact that there was no monthly consistency in the timing of their deposits or withdrawals.

One of the most graphic examples of large-scale data use is a recent advance in the computer translation of languages. Using a computer to translate between one language and another has been an aim of researchers since the 1960s. The normal approach to translate from a language *A* into a language *B* has been to devise a set of rules which describe the sentence structure of language *A*, process the text in language *A* that needs to be translated, discern its structure using the rules, and then using the rules of sentence structure of language *B* transform the text into its equivalent in that language.

Progress in what is known as 'machine translation' has been steady but slow. However, recent work at Google has accelerated the process. Researchers at the company have developed a new technique which relies on the use of a huge corpus of existing text. The researchers used a technique known as 'machine learning' to carry out the translation. Machine learning is a means by which a computer program is given a large number of cases and results and tries to infer some hypothesis from the data presented. In the Google case, the machine-learning program was presented with 200 billion words of text and their equivalents in other languages that were supplied by the United Nations; it then learned the way in which individual sentences were rendered from one language to another, without having any knowledge of the structure of any of the languages used. The next few years hold out the promise of improvement compared with programs that use rule descriptions of a language.

In his book *Planet Google*, Randall Stross describes how an Arabic sentence was transformed by a commercial translation system into 'Alpine white new presence tape registered for coffee confirms Laden', when the Google system translated it almost perfectly into 'The White House confirmed the existence of a new Bin Laden tape'.

At the start of this chapter, I described how loyalty cards provide useful information to retailers such as Tesco. This is an example of developing a database that can be used for a variety of marketing purposes and employing a computer to process it. There are many more. One major application of data-mining is market-basket analysis. Here, a company will keep past purchases and other data such as demographic information in order to improve its sales. For example, Amazon uses past sales to notify customers of items that they may find interesting and which are somehow associated with their past sales; for example, a customer who buys crime novels will almost invariably be presented with newly published crime fiction when they log on to the Amazon web site again.

Another example of the use of market-basket analysis is that of identifying what are known as 'alpha customers'. These are customers who have a career that places them in front of a large number of people. They may be celebrities who are often seen on television, and whose images are found in newspapers, or they may be business leaders who will be listened to in seminars and shareholder meetings. What is important is that these are people whose lifestyles are often copied or their advice taken.

Such an application of data-mining is only made profitable by a concept known as viral marketing whereby social sites such as Facebook amplify conversations between consumers and spread these conversations. There are a number of examples of how successful this can be. The film *The Blair Witch Project* would, in pre-Internet days, have become not just a cult film but a minor cult film; electronic word of mouth made it hugely successful.

One interesting commercial use of the technique of aggregating data is that of the prediction market. This is similar in many ways to a stock market where investors buy and sell shares in companies. However, in a prediction market it is a prediction: it might be a prediction about who will win an Oscar, what the election majority of a political party will be, or economic events such as a major change in a currency rate. When such markets are popular they can predict to a high degree of accuracy.
For example, a market that involved the buying and selling of

Oscar predictions correctly predicted 32 of the 36 big category nominees for the 2006 Oscars, and 7 out of the 8 top category winners.

From an Internet phenomenon prediction markets have now become something major companies use for economic prediction. For example, companies such as Google use internal prediction markets to determine business policy.

At the beginning of this book, I defined the computer as:

> A computer contains one or more processors which operate on data. The processor(s) are connected to data storage. The intentions of a human operator are conveyed to the computer via a number of input devices. The result of any computation carried out by the processor(s) will be shown on a number of display devices.

You may have thought that this was the academic in me coming out: relying on a definition, tying up loose ends, and attempting to be semantically clear. However, the point this definition makes is that it covers not only the computer on your desk that you use for tasks such as word-processing and web browsing, the computers that help fly a plane, or the computer chip embedded in an iPod, but also a much larger computer made up of hundreds of millions of computers and which is embedded in a cloud; I hope that this chapter has convinced you of this.

Chapter 8
The next computer

Introduction

The basic architecture of the computer has remained unchanged for six decades since IBM developed the first mainframe computers. It consists of a processor that reads software instructions one by one and executes them. Each instruction will result in data being processed, for example by being added together; and data being stored in the main memory of the computer or being stored on some file-storage medium; or being sent to the Internet or to another computer. This is what is known as the von Neumann architecture; it was named after John von Neumann, a naturalized American mathematician who was at the forefront of computer research in the 1940s and 1950s. His key idea, which still holds sway today, is that in a computer the data and the program are both stored in the computer's memory in the same address space.

There have been few challenges to the von Neumann architecture. In this final chapter, I shall look at two future approaches to computing which involve totally different architectures. I shall also look at a sub-architecture which is known as the 'neural net architecture', which has some resemblance to the neural structures in the brain. But, first, a strange piece of history which may repeat itself.

Functional programming and non-standard architectures

In the 1980s, there was a huge increase of research funding for computing researchers – undreamt of funds became available from British, European, and American sources. It would be nice to think that this occurred because of governments realizing the potential of computers. However, it occurred because of fear: fear of Japanese industries. The Japanese ministry of trade, MITI, had announced a major programme of funding into computer technology and Western governments, having experienced the havoc that the Japanese electronics and motor industries had caused, feared similar things might happen to their still-evolving computer industries.

The United Kingdom set up the Alvey programme. This was a joint programme between government departments and the Science Research Council, to advance both research and the supply of trained IT staff. There were a number of strands to this research, one of which was novel computer architectures.

In the 1980s, a number of researchers had described problems with software development and with the von Neumann architecture. Some pointed out that software developed with conventional programming languages was often error-ridden and that the painstaking process of instructing the computer to read data, process it, and write to memory was too detailed and complex. Other researchers pointed out that in the coming decades, hardware technology would advance to the point where large numbers of processors could be embedded on a single chip: that this offered major opportunities in terms of computing power, but would create more errors as programmers tried to share work amongst the processors.

In order to address these problems, computer scientists started developing a new class of programming languages known as

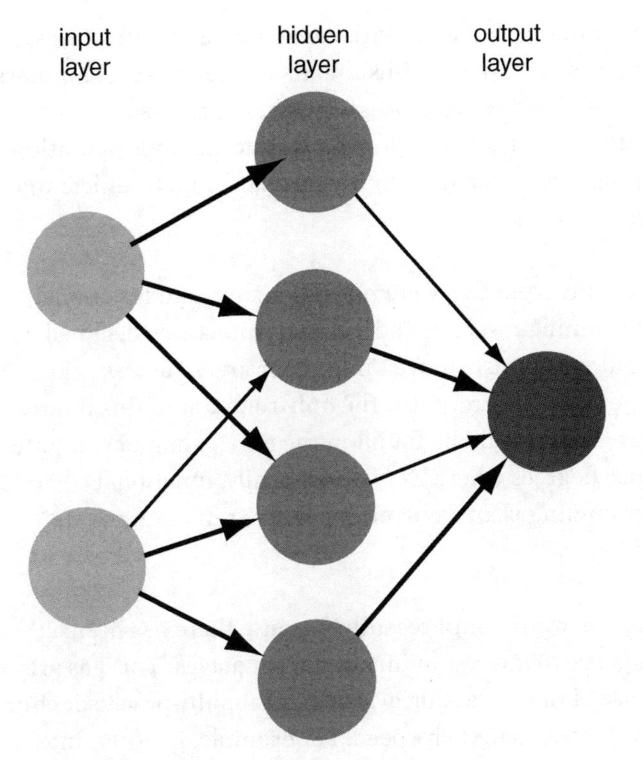

9. A simple neural network

'functional languages'. These languages did not have the step-by-step feature of a conventional programming language; rather, they consisted of a series of mathematical equations that defined the end result of the computation. It was left to the computer to 'solve' these equations and deliver the correct result.

A number of functional languages were developed: the four best-known are FP, Haskell, ML, and Miranda. A number of computer packages were also developed to execute programs in these languages. Unfortunately, there was a problem: the programs resided on von Neumann computers; this resulted in very slow execution times – conventional architectures were just not up to the job.

Because of the problems with conventional architectures, a number of research establishments created novel computers out of arrays of hardware processors. These processors would take a functional program and process an internal representation of the program, reducing it to a form that could be efficiently executed.

The results from this were not promising, and functional programming research and novel architectures declined as computer scientists moved into other areas such as grid computing. Until recently, the only remnant of this interesting research area could be found in the curriculum of computer science degrees where, very occasionally, functional programming is an element in first-year undergraduate teaching.

I used the words 'until recently' because there has been a resurgence of interest in functional languages. This has arisen because of the increasing availability of multi-processor chips and major increases in their speed. For example, in 2009, Intel announced a processor, the Single-chip Cloud Computer, that had 48 processors on a piece of silicon no bigger than a postage stamp. It has meant that, for example, publishers who are mainly associated with professional sales rather than academic sales have been releasing books on functional programming languages. As desktop computers become more powerful and include more processors, such languages should achieve much greater penetration, even if they are implemented on conventional architectures.

An interesting architecture that has applications over a relatively limited domain is the neural network. It is an architecture that has often been simulated on a von Neumann computer. It is used for pattern-recognition applications and for predicting short-term trends.

A cancer cell detection system based on neural nets was developed by Andrea Dawson of the University of Rochester Medical Center, Richard Austin of the University of California at San Francisco, and David Weinberg of the Brigham and Women's Hospital; this system can detect cancerous cells with an accuracy comparable to human detection.

Researchers at Queen Mary College, part of the University of London, have used neural networks in a security application where they are used to focus on the face of a possible intruder in a building even though the intruder may be moving.

Researchers at the University of Sussex have used the genetic programming idea that I described in Chapter 4 to create a number of generations of neural computers and select the best neural computer that solves problems in the area of pursuit and evasion where someone or something is trying to avoid a pursuer.

So, what is a neural computer, often called a 'neural network'? Initially, the idea of such a computer arose from studies of the neural architecture of the brain. Figure 9 shows a simple schematic. It consists of an input layer that can sense various signals from some environment, for example the pixels that make up a picture of a human face. In the middle (hidden layer), there are a large number of processing elements (neurones) which are arranged into sub-layers. Finally, there is an output layer which provides a result, for example this might be a simple window on a security system when, say, an airline passenger who is a potential terrorist is recognized.

It is in the middle layer that the work is done in a neural computer. What happens is that the network is trained by giving it examples of the trend or item that is to be recognized. What the training does is to strengthen or weaken the connections between the processing elements in the middle layer until, when combined, they produce a strong signal when a new case is presented to them that matches the previously trained examples and a weak signal when an item that does not match the examples is encountered.

Neural networks have been implemented in hardware, but most of the implementations have been via software where the middle layer has been implemented in chunks of code that carry out the learning process.

Before looking at other novel architectures, it is worth saying that although the initial impetus was to use ideas in neurobiology to develop neural architectures based on a consideration of processes in the brain, there is little resemblance between the internal data and software now used in commercial implementations and the human brain.

Quantum computers

In Chapter 4, I discussed how VLSI technology has enabled computer companies to develop more and more powerful computers; the increase in power arising from an increased ability to squeeze more and more processors on a single chip.

The von Neumann computer is based on the storage of data using binary digits (0 or 1) collected together into bytes. Quantum computers store data as quantum bits, or qubits; these can be implemented as atoms, ions, photons, or electrons. These qubits can not only act as a storage element for data but can also be combined together to implement a hardware processor.

In order to get an idea of how powerful a quantum computer can be, David Deutsch, an Oxford academic and physicist, and one of the pioneers in the area, has calculated that a modest 30-qubit computer could work at 10 teraflops. This is comparable to the speeds achieved by the supercomputers that were working in the first decade of this millennium.

One of the problems with using quantum ideas to implement a computer is concerned with the effect that observation of qubits could have. For example, trying to examine the state of a qubit could change its state along with others. This means that it could be immensely difficult to read data from a quantum computer – a process that is very easy with a conventional computer.

Happily, there is a quantum phenomenon known as entanglement which has come to the aid of the quantum computer developer. In 2006, physicists at the USA Commerce Department's National Institute of Standards and Technology reported in the 19 October edition of the science journal *Nature* that they had taken a major step towards moulding entanglement into a technique that can be used to read quantum-based data. They demonstrated a method for refining entangled atom pairs – a process called purification – which involved entangling two pairs of beryllium ions. This means that the data in a quantum computer can be observed indirectly without affecting its value.

Quantum computing is in its very early days. Results have been very small-scale so far: for example, a Canadian startup company, D-Wave, has demonstrated a working 16-qubit quantum computer. The computer solved a simple Sudoku puzzle and other pattern-matching problems. Now, compared with the performance of a conventional computer, this solution of a modestly sized problem is no huge achievement in terms of capability. However, it is a dramatic proof of concept.

Quantum computers are important in that a successful computer based on quantum physics ideas could overturn many of the technologies that are currently extant. One of the areas where it would have the most effect is in cryptography. A number of modern cryptographic methods rely on the fact that it is exceptionally hard to solve certain problems – problems that are known as wicked problems – however, researchers have pointed out that a relatively modest quantum computer *could* solve these problems; indeed, one of the challenges that this community have addressed is of doing just that.

For example, in 2001 computer scientists from IBM and Stanford University demonstrated that a quantum computer can be programmed to find the prime factors of numbers (a prime factor is a number which divides exactly into a number and which cannot be reduced further; for example, the prime factors of 33 are 3 and 11). The researchers employed a 7-qubit computer to find the factors of 15 (5 and 3). Prime factor determination is one of the problems that enable Internet-based cryptography to be successful.

Again, in computation terms this is no great achievement; in 2005, for example, researchers at the German Federal Agency for Information Technology Security used a cluster of conventional computers to factor a 200-digit number. However, it represents an important proof of concept which, if made industrial, would threaten a large segment of the technologies we use to send secure data.

The DNA computer

Deoxyribonucleic acid (DNA) is a nucleic acid that contains the genetic instructions used to determine the development and life of living organisms and some viruses. When biologists refer to DNA, they often talk about it in terms of a computer program: a program that, when executed, controls our physical growth and that makes up our genetic profile.

DNA contains data that consists of two long polymers of simple units called nucleotides. Within the DNA there are also sugars and phosphates. Attached to each sugar is one of four categories (A, T, C, and G) of molecules that are known as bases. These are the basic units that encode the information used in our genetic development.

Thus, we have a potential computer containing data and a program that manipulates this data. As early as 1994, the American computer scientist Leonard Adelman showed that a computer based on DNA could solve a problem known as the directed Hamiltonian path problem. This is a special version of the travelling salesman problem that was described in Chapter 4. The number of data points that were processed by Adelman's program were small: a desktop computer could have obtained the correct solution in a short amount of time. However, it represented a proof of concept.

These are the steps that this DNA computer might carry out in order to solve a problem: the strands of DNA would be regarded as the data for the problem. These would be codified using the A, T, C, and G encoding found in the bases.

These molecules would then be mixed, and a number of these DNA strands would stick together. Each collection of strands would represent a possible solution to the problem. Invalid answers would be removed by a series of chemical reactions. What would be left would be one or more solutions to the problem being solved.

After Adelman's epoch-making work, a number of research groups, alerted to the potential of using human genetic material, started looking at analogues between DNA and the von Neumann computer. Researchers at the University of Rochester have developed electronic circuits known as 'logic gates' out of genetic material. As you may remember from Chapter 1, a logic gate is an electronic device that takes a number of bits and produces as an output another binary bit that depends on the inputs. For example, And logic gates produce a 1 when their two inputs are 1 and a zero otherwise. In the same way that the components of DNA are the fundamental building blocks of life, the gates are the fundamental building blocks of computers; for example, they are used to implement arithmetic functions such as addition and subtraction.

DNA computers are clearly a radical departure from current computers in that the material used to produce them is of a chemical nature – current computers are made out of metal and silicon. However, they do not depart from the von Neumann idea, as the quantum computer does. What they do offer is a computation paradigm that offers massive amounts of storage and very little power dissipation. Like the quantum computer, they are only just emerging from the proof of concept stage.

A view from dystopia

Jonathan Zittrain is a distinguished Professor of Internet Governance. He has written a key work on the future of the Internet – not the technical future, but a future that could be constrained by government and industrial actions. It might seem odd to finish this book with a chapter that represents a somewhat dystopic view of the computer; after all, this entire book has been about the computer and has concentrated on the staggering advances it has made – so why finish with the Internet, and why finish on a downbeat note? The first question is easy to answer: this book has continually stressed that you cannot disengage the computer from the networked environment within which it now operates. The answer to the second question is that it is too easy to get carried away with the gee-whizzery of computing without considering some of the down points. I have done this a little in Chapter 3, where I briefly explored some issues of security and privacy.

The idea that the computer can solve major problems without, for example, looking at context, is an idea that still seems to have a great deal of currency. An example of this is the wastage that is associated with British government IT projects; large projects, such as the National Health IT project, have leaked money over the last decade. For example, a recent report by the Taxpayers' Alliance documented a large number of government projects that had reported budget overruns. It detailed overruns of around £19 billion, of which IT-based projects contributed £12 billion.

In the first part of his book, Zittrain describes the history of the computer. It starts at the point when individual computers were maintained by a professional class and when software systems satisfied narrow requirements, for example to produce the wage and salary slips of employees – this occurred around the 1960s. He then describes the processes whereby the PC gradually became part of many people's electronic furniture so that the non-technical user could install whatever software they wished, to the point when, via the Internet, it became a tool for mass creativity and mass collaboration.

In his book, Zittrain looks at the tensions between aspects of this metamorphosis: declining security, as detailed in Chapter 3, declining privacy as detailed in Chapter 5, and the notions of the computer and the Internet as a public property capable of informing, enabling collaboration, and encouraging creativity.

Zittrain describes a number of possible brakes on the development of the computer in terms of what it might be possible to do with it in the future. One potential brake is that of devices that allow access to the Internet, only in a way controlled by the manufacturers of the device. These devices include game consoles, MP3 players, digital TV recorders, and e-readers.

Such devices are either completely closed or almost fully closed and users are only allowed access and modification of the underlying software under strict conditions – usually such users develop third-party software for the device and are allowed access only after some commercial agreement is made.

A possible future implied by Zittrain is of an Internet in which access is via home computer terminals with little of the hardware power of the current PCs – they would just connect to the Internet – and where a host of tethered devices restrict the user in the same way that, for example, the North Korean government restricts external wireless reception in their country.

Zittrain's theme was preceded by ideas propanded by Nicholas Carr in his book *The Big Switch*. Carr's focus is commercial. In the book, he compares the growth of the Internet to the growth of electrical power generation. In the latter part of the 19th century and early part of the last century, power generation was carried out almost as a cottage industry, with each company maintaining its own generator and employing heavy electrical engineers to maintain them. Carr's book shows how Thomas Edison and Samuel Insull revolutionized the electricity-generation industry by developing an infrastructure whereby, initially, domestic consumers and then industrial consumers took their power from a grid. The move to electricity was rapid and was the result of advances in power generation and the radical step of replacing direct current power with alternating current power.

Carr compares this growth of electricity generation with the growth of the Internet. As you will have seen in Chapter 7, there is an increasing trend for computer power to be centralized. For example, there is an increasing number of sites that store your data securely; the grid allows the Internet to take the processing strain away from the PC; and, increasingly, companies are offering software functionality via a web site rather than by providing a facility where software can be installed on a PC or a server.

The infrastructure is already there for the replacement of the PC by cut-down, computer-based consumer devices that feature a screen, a keyboard, and enough hardware to allow the user to communicate with the Internet together with unmodifiable software embedded in firmware.

Carr's thesis is also supported by another writer, Tim Wu. In his book *The Master Switch*, Wu also takes a historical perspective which is based on the growth of television and other communication media. He points out that those technologies that were once free and open have eventually become centralized and closed as a result of commercial pressures.

So, another image of the computer of the future is as an embedded device within a piece of consumer electronics that interacts with the Internet via a browser; in which activities such as developing a piece of text would be carried out using a word processor held on a remote server. It would be a piece of electronics that prevents the user from modifying the device via software. In a few decades time, we may look at the desktop or laptop we currently use in the same way we look at photographs of early electricity-generation hardware.

Further reading

In this section, I have included a list of excellent books that are relevant to each chapter and which expand upon the main points made in each. Some of these are relevant to more than one chapter, so you will find some duplication.

The naked computer

M. Hally, *Electronic Brains: Stories from the Dawn of the Computer Age* (Granta Books, 2006).
C. Petzold, *Code: the Hidden Language* (Microsoft Press, 2000).

The small computer

C. Maxfield, *BeBOP to the Boolean Boogie: An Unconventional Guide to Electronics* (Newnes, 2008).

The ubiquitous computer

N. Christakis and J. Fowler, *Connected* (HarperCollins, 2011).
A. Greenfield, *Everyware: The Dawning Age of Ubiquitous Computing* (Peachpit Press, 2006).
J. Krumm (ed.), *Ubiquitous Computing Fundamentals* (Chapman and Hall, 2009).
K. O'Hara and N. Shadbolt, *The Spy in the Coffee Machine* (Oneworld Publications, 2008).

The global computer

C. J. Murray, *The Supermen: The Story of Seymour Cray and the Technical Wizards Behind the Supercomputer* (Wiley, 1997).

J. Naughton, *A Brief History of the Future: Origins of the Internet* (Phoenix, 2000).

M. M. Waldrop, *The Dream Machine: J. C. R. Licklider and the Revolution That Made Computing Personal* (Penguin Books, 2002).

J. Zittrain, *The Future of the Internet* (Penguin Books, 2009).

The insecure computer

K. O'Hara and N. Shadbolt, *The Spy in the Coffee Machine* (Oneworld Publications, 2008).

K. D. Mitnick and W. L. Simon, *The Art of Deception* (Wiley, 2003).

C. P. Pfleeger and S. L. Pfleeger, *Security in Computing* (Prentice Hall, 2006).

F. Piper and S. Murphy, *Cryptography: A Very Short Introduction* (Oxford University Press, 2002).

B. Schneier, *Secrets and Lies* (Wiley, 2004).

B. Schneier, *Beyond Fear: Thinking Sensibly about Security in an Uncertain World* (Springer, 2003).

The disruptive computer

C. M. Christensen, *The Innovator's Dilemma: When New Technologies Cause Great Firms to Fail* (Harvard Business School Press, 1997).

C. M. Christenson and M. E. Raynor, *The Innovator's Solution: Creating and Sustaining Successful Growth* (Harvard Business School Press, 2003).

J. Gomez, *Print Is Dead* (Macmillan, 2007).

D. Silverman, *Typo: The Last American Typesetter or How I Made and Lost 4 Million Dollars* (Soft Skull Press, 2008).

The cloud computer

I. Ayres, *Super Crunchers* (John Murray, 2008).

N. G. Carr, *Does IT Matter? Information Technology and the Corrosion of Competitive Advantage* (Harvard Business School Press, 2004).

N. G. Carr, *The Big Switch: Rewiring the World, from 'Edison' to 'Google'* (W. W. Norton, 2009).

B. Mezrich, *The Accidental Billionaires: The Founding of Facebook – A Tale of Sex, Money, Genius and Betrayal* (Doubleday, 2010).
R. E. Stross, *Planet Google* (Atlantic Books, 2009).
B. Tancer, *Click* (Hyperion, 2009).
D. Tapscott and A. D. Williams, *Wikinomics* (Atlantic Books, 2006).

The next computer

N. G. Carr, *The Big Switch: Rewiring the World, from 'Edison' to 'Google'* (W. W. Norton, 2009).
J. Gleick, *The Information: A History, a Theory, a Flood* (Fourth Estate, 2011).
T. Wu, *The Master Switch* (Atlantic Books, 2011).
J. Zittrain, *The Future of the Internet: And How to Stop It* (Penguin Books, 2009).

“牛津通识读本”已出书目

古典哲学的趣味
人生的意义
文学理论入门
大众经济学
历史之源
设计，无处不在
生活中的心理学
政治的历史与边界
哲学的思与惑
资本主义
美国总统制
海德格尔
我们时代的伦理学
卡夫卡是谁
考古学的过去与未来
天文学简史
社会学的意识
康德
尼采
亚里士多德的世界
西方艺术新论
全球化面面观
简明逻辑学
法哲学：价值与事实
政治哲学与幸福根基
选择理论
后殖民主义与世界格局

福柯
缤纷的语言学
达达和超现实主义
佛学概论
维特根斯坦与哲学
科学哲学
印度哲学祛魅
克尔凯郭尔
科学革命
广告
数学
叔本华
笛卡尔
基督教神学
犹太人与犹太教
现代日本
罗兰·巴特
马基雅维里
全球经济史
进化
性存在
量子理论
牛顿新传
国际移民
哈贝马斯
医学伦理
黑格尔

地球
记忆
法律
中国文学
托克维尔
休谟
分子
法国大革命
民族主义
科幻作品
罗素
美国政党与选举
美国最高法院
纪录片
大萧条与罗斯福新政
领导力
无神论
罗马共和国
美国国会
民主
英格兰文学
现代主义
网络
自闭症
德里达
浪漫主义
批判理论

德国文学
戏剧
腐败
医事法
癌症
植物
法语文学
微观经济学
湖泊
拜占庭
司法心理学
发展
农业
特洛伊战争
巴比伦尼亚
河流
战争与技术
品牌学
数学简史
物理学
行为经济学
计算机科学
儿童心理学
时装
现代拉丁美洲文学
卢梭
隐私
电影音乐
抑郁症
传染病
希腊化时代
知识
环境伦理学
美国革命
元素周期表
人口学
社会心理学
动物
项目管理
美学
管理学
卫星
国际法
计算机
电影
俄罗斯文学
古典文学
大数据
洛克
幸福
免疫系统
银行学
景观设计学
神圣罗马帝国
大流行病
亚历山大大帝
气候
第二次世界大战
中世纪
工业革命
传记
公共管理
社会语言学
物质
学习